Science Matters
Ben Wright

"We live in a society absolutely dependent on science and technology and yet have cleverly arranged things so that almost no one understands science and technology."
Carl Sagan

Introduction

Have you ever wished that you could better understand science? Yes? Then maybe I can help you a little. More than likely if you have opened up this book then you have at least some interest in science? Science is without a doubt human kinds greatest ever accomplishment and something that everyone should be able to understand as much as they enjoy it accomplishments. Great scientific discoveries have been the main focus of our species since we evolved the ability to think, and this ability has become our primary survival tool.

The aim of the book is to try and encourage the public perception of science by improving the understanding of science. I hope this book will show people that science doesn't have to be all beards and glasses, whiteboards and microscopes. 'Science Matters' aims to show people that science is not as confusing or boring as you might have been led to believe, and that science subjects (as well as scientists) are not as inaccessible as you might think. Incidentally, I don't wear glasses and I can't grow a beard. I have more than twenty tattoos (I've lost count) and I go to rock concerts when I'm not at the gym or at rugby training.

Science is exciting and science is fun. Scientists all over the world are people who are making all the greatest discoveries. I think that being a scientist is certainly a lot more glamorous, exciting and fun than being a hairdresser or a builder, or working in an office.

How can it be that scientists can be given such a hard time by pop culture? Science is what makes the world go round, and a sound grasp of scientific knowledge can make you feel fulfilled in a way that nothing else can. Science is the greatest creation of the human mind and a wonderful thing, unless of course, as Albert Einstein once said: '... one does not have to earn one's living at it.'

This book gives you plenty to think about. There are some chapters that might make you think, "wow, I've always wondered about that" and some that might make you say, "I would have never wondered about that but it was still interesting."

The chapter topics are diverse and pretty random and written with a unique voice that intend to make each one enjoyable and easy to follow, making sure you never feel your reading a text book. You can pick a chapter or two from the list, or read through them all. But in any case, I hope you enjoy it!

The public perception of Science

A great man once said, "science is interesting, and if you don't like it, you can f*** off!" he was the first editor of New Scientist magazine, and his outburst was intended in sarcastic jest. However, there is a somewhat serious undercurrent to his proclamation. To the majority of the general public, science is of little or no interest.

Science's poor public image is widespread amongst the general public. When people think of science they think of personalities like Einstein and Stephen Hawking. They might think of be-spectacled, middle-aged eccentrics working excitedly in a laboratory. Worst of all thinking of science may bring back memories of being in school, bored and desperately waiting for the bell to ring, signalling the end the class. All of these are fairly accurate depictions of science. Einstein and Hawking are very famous scientists, many scientists are middle-aged, work in labs and wear glasses and science teaching in schools does not generally make the most of the material.

Science suffers an image problem, it's impossible to deny, but does it have to be that way?

You're a geek mate!
Certainly amongst many of my personal acquaintances, it is seen as almost an admirable quality to be ignorant of science, something that would never be admitted if it were in reference to cars or football or similar.

I remember on a particular bus trip to play a rugby match, I dared to read a book on the way.

"What you reading?" I was asked
"Just a book" I replied tentatively
"What's it about?" I was then asked, a question I have now come to resent,
"Oh...Em...It's about Quantum Physics" I answered, knowing what the next question would be,
"Oh right...great" he said sarcastically while walking away.

I was perplexed. I had expected him to ask me to explain quantum physics (which I was not looking forward to doing) but instead nothing but a supercilious departing comment.

What is wrong with science?

So, why is it perceived as okay to be ignorant of science? Why, in some individuals, is it considered negative or even reprehensible to have an active interested in science?

It may be largely to do with the negative stereotypes attached to science and scientists. Most people envisage the scientist as an awkward, anti-social, oddball. Not exactly an inspirational, luminary figurehead that most people can relate or aspire towards.

Scientists are very rarely public personalities, they very rarely seek notoriety (out-with the scientific community) and many actually actively abnegate celebrity. As a result, science is not an inspirational occupation for most.

Any child who announced at age 4 "I want to be a palaeontologist when I grow up" would be considered strange (that was me by the way). As a result of this negative PR most people just don't think that science is for them. It's too complicated, or it's too hard or it's just not interesting.

Science is beautiful

The reality is that science is exciting; it is inspirational and can be beautiful. The romanticist poet John Keats once accused Isaac Newton of destroying the poetry of the rainbow by explaining the origins of its colours.

"Do not all charms fly, at the mere touch of cold philosophy? …"

Keats could not have been more wrong. Science has led us to discover some of the most beautiful phenomena imaginable.
Science brings us all closer to these stunning majesties and continues to do so through many small steps of increasing knowledge, slowly building towards a consensus of wonder. Science is by far the most beautiful, inspirational and life altering creation of the human mind. When talking about science, Richard Feynman once said "You gotta stop and think about the complexity to really get the pleasure" and maybe this is the problem. To really 'get it' and fully understand the beauty and poetry in science you have to have an understanding of science.

For example, one of the most beautiful concepts of science is that we are connected to everything in the entire universe. We are connected to each other, biologically. We are connected to the earth, chemically. We are connected to the entire universe, atomically. This sounds like a hugely unlikely statement, yet with even a slight grasp of science you would know that this was indeed true. As Carl Sagan once said we indeed "made of star stuff"

If only people would let science into their lives, I believe countless lives would be greatly improved, for the imagination of science's inconceivable nature is so much greater than mans could ever be.

Why we are lucky to be alive

"Know now that you are born along with these
clouds, winds, and stars, and ever-moving seas
and forest dwellers. This your nature is.

Lift up your heart again without fear,
Sleep in the tomb, or breathe the living air,
This world you with the flower and the tiger share."

These two brief verses are taken from the poem "Passion" written in 1943 by Kathleen Raine. I first read the poem in a Richard Dawkins book and it left quite an impression on me.

"So what!" you might be thinking. "What has this got to do with science?" Well actually, it has EVERYTHING to do with science. Not that "Passion" is a poem about science as it isn't, although it could be. The poem is an ode to the wonders of existence, and how we are all very lucky to be part of such and incredible and beautiful world full of astounding natural phenomena. This is where the poem really impacts on me emotionally and makes me feel so lucky to share this moment in the spotlight of time in such unique and remarkable world. Ok so maybe I'm getting a little bit carried away.

Oscar Wilde was famously quoted, "To live is the rarest thing in the world. Most people exist. That is all." The world renowned Irish playwright, poet and author in Oscar Fingal O'Flahertie Wills Wilde's typical witty and biting philosophy, was merely to drawn attention to how many people fail to fulfil their lives potential, and was not to be taken literally.

That said, we are literally very lucky to be alive and that is what I shall talk about for the remainder of this chapter. I'm going to take a slight respite from the straight science and be a little bit more philosophical, although everything I will say is scientifically accurate of course. It has been said that philosophers are like tourists, whereas scientists are explorers, but I digress.

Potential people
To borrow again from Professor Dawkins, "we are all going to die, and that makes us the lucky ones". What may sound like a confusing or even paradoxical statement is indeed neither. If one is lucky enough to die it

also means that they have been lucky enough to have lived. As the number of "potential people" who never made it to be lucky enough to become living people vastly out numbers the individual raindrops to have ever struck a single patch of forest canopy.

Before we move onto the unbelievable lottery of our own conception, we must first realise that the incredible singularity of birth began far before any of us were conceived. Firstly, of course for our parent's individual DNA to meet and create our personal DNA, they had to reproduce. Before this could occur of course they had to meet, by coincidence more than likely and date for a while, get along with each other (maybe) and product offspring. However, to make each of your parents, their parents had to meet and mix DNA in the same manner and their parents similarly and indeed their parents as well. In fact, the specific DNA mixture that makes you as an individual is directly dependant on an enormous lineage of reproduction continuing right back to the first life ever to have evolved on this planet. This might sound far-fetched but you, me and everything else on this planet is related to each other. We are all connected through an unbroken line of successfully reproducing offspring right back to the dawn of life. That is a lot of mixing of DNA that has gone together to make YOU. So how lucky are YOU to be where you are?

Sperm wars
The biggest lottery of your personal existence however occurs at conception. Nine months before your birth there was an extremely important event to your personal fortunes. It was the moment in which your chances of coming into existence became hundreds of billions of times more likely and it was the millisecond before your conception. Each man produces millions of sperm, only one contains YOUR specific DNA offering and only one will make it to become a person. Anyone who has read the book "sperm wars" or indeed has watched "look who's talking" knows that it is something akin to a battlefield for the sperm to even reach the egg and even once this has occurred the embryonic you is still far from safe. Most conceptions are actually aborted by the body before the mother has any knowledge that it had occurred. So yet again, we were all very lucky to have dodged that bullet. However, as we all were lucky enough to, the chances of becoming a walking, talking, living human have drop from astonishingly small to quite likely. Now all that need to be done is survive pregnancy and birth, which most babies nowadays fortunately do.

Conclusion ... perfect planet

You can actually take the luckiness back even further which makes you feel even more fortunate to be alive. When you consider how precise and particular the conditions that are needed for life to even exist on earth. So particular that so far we have not found a single other planet with any evidence of life, even extinct life. Just take a quick glimpse at Earth, not too hot, not too cold, only gently spinning, oxygen to breath, water to drink, green and fertile, bathed in sunshine. How lucky we are, not only that we are alive but that we have evolved in a way that we are able to appreciate just how lucky we are to be living.

What makes us fall?

"Gravity is full of worth
Holds me nicely down to earth.
Without it, my hair would tend
to stand up straight on its end."

Rusty Daily (internet poet)

I don't know if you have ever given the act of falling over much more thought than "Oh Sh** I'm gonna fall" or "ouch, that hurt" but the reason that we fall over is one of the most important and interesting of any of the forces in the entire universe for somebody to ever wonder about.

Gravity, which of course is what makes things appear to fall, has a lot of other vital roles on our planet, all of which are of special importance to us. If it were not for gravity there would be no clouds, no rain, no rivers, no oceans, no wind, no walking, no plants, no haircuts, no aircraft, no eating, no drinking, no going to the toilet, no sleeping, no writing, no swimming, no fire, no life. I'm going to stop actually because I could go on forever like this and I think you understand my point, without gravity there would not be much of anything because everything is reliant on it.

So what is gravity?
One of the two major natural forces (the other being electro-magnetism) which we experience every day, gravity is truly remarkable. It seems like an extremely simple and easily understood force of nature, simply the force which holds all things towards the earth. When you look at it from the point of view of a physicist however, gravity becomes a little more complex to understand and I suggest (hopefully without sounding patronising) that you read the rest of this article slowly, just for that reason.

Gravity = $(G*m1*m2) / (d2)$

In this 'simple' equation, **G** represents the gravitational constant, **m1** and **m2** represent the mass of the objects for which you are calculating the force, and **d** represents the distance between the centers of gravity of the two masses.

Does that mean anything to you because it certainly doesn't mean much to me? The aforementioned formula was the first formula to explain gravity and was devised by Isaac Newton. It is also the simplest explanation and has, in reality been superseded by Albert Einstein's MUCH more complex explanation. Which is going to be the primary focus of this article, I am going to try and explain how gravity is currently understood to work in simple terms. Gravity is a simple function, the reality of which is unfortunately extremely difficult to understand. Even the most basic explanation of how gravity works has never really satisfied me and always leaves me wishing I was more intelligent. I am going to try however, even if it involves completely ignoring a lot of the detail...I won't say anything if you don't.

Newton's ideas about gravity

As I'm sure most people are familiar, as legend has it the English mathematician Sir Isaac Newton was the first to satisfactorily explain the force of gravity in scientific terms in 1687, after allegedly being struck on the head by a falling apple. In reality the aforementioned apple incident probably never happened, and Newton was not actually the first to explain the theory of Gravity. 42 years earlier Ismael Bullialdus had put forward an almost identical theory, which for some reason did not take off then.

Newton's theory stated that one mass is attracted to another mass by a force which is proportional to the product of the two masses and inversely (oppositely) proportional to the distance between them, multiplied by itself. The gravitational constant is the measurement of the gravitational attraction between two objects with mass and is the most important part of Newton's theory. So put simply, Newton said that gravity is a measurement of the force of attraction between two objects with mass, related to the distance between them. Or put even more simply, gravity is the force pulling together all things in the universe.

Newton's explanation of gravity was (and still is) an extremely successful one and should not be considered for a second to be wrong, as it is NOT. Newton's theory has enjoyed many great triumphs, all of which proved it to be true and accurate. The greatest of these was probably its success in predicting the existence of the planet Neptune and the movements of the planet Uranus.

One of the most important roles of scientific research is its ability to be tested to the extent that it will point out any flaws in current ideas, which makes the understanding of the theories stronger and our knowledge greater. This is what happened with Newton's theory of gravity, it was actually through the use of Newton's theory that scientist noticed that it didn't quite explain what they were able to see in the orbit of the planet Mercury. This issue was then resolved, and Newton's theory was superseded by Einstein's General Theory of Relativity...which is going to be great fun to try and explain. Before I move on to try and do that, it is important to realize that Newton's theory has not been totally superseded and is still more commonly used than Einstein's theory, due to that fact that it is much simpler to work with and is accurate enough for most calculations.

Einstein's ideas about gravity

Albert Einstein understood that Newton's theory had a few small weaknesses and solved these problems by looking at gravity from a completely different point of view from anybody else. In doing so Einstein produced what is considered by many to be the greatest and most profoundly brilliant individual pieces of science ever produced. In General Relativity the effects of gravity are not explained by any type of force (as they are by Newton). Instead they are described as a consequence of the fact that space-time is not flat, but curved (or warped) by the distribution of mass and energy within it.

What Einstein meant was that gravity is actually a property of the geometry of space and of time. Specifically, he described how the curvature of space-time is directly related to the momentum of any matter and radiation which is present. Einstein said that space-time is curved by matter and that falling objects move along straight paths in curved space-time. Try and imagine that things such as the planets are not made to move on curved orbits by a force but instead they move in straight lines through a curved space (called a geodesic). In general relativity bodies only move in straight lines through four-dimensional space-time; but as we live in a three-dimensional world they appear to move along curved paths, and this acceleration produced is what is called gravitation. So, in reality the curved orbits of the planets are actually straight paths towards the sun, which is causing the curvature of space-time and the illusion of orbits around the sun. That explains why planets such as Mercury which is the closest to the sun and therefore experiences the greatest gravitational pull, also experience the most exaggerated orbit.

I'm not going to actually try and explain how General Relativity actually works, as it is far too complicated and I don't understand it. What I do however understand is how General Relativity has enjoyed a great deal of success due to the fact that the predictions which it has successfully made, which Newton's theory of gravity could not explain, have been confirmed. In most cases they exactly match the Newtonian predictions and in other cases they explain things Newton could not.

"Nature, and Nature's laws lay hid in night:
God said: '*Let Newton be!*' and all was light.
It did not last: the Devil howling '*Ho!*
Let Einstein be l*' restored the status quo."

The "complete" theory?

As I have now explained and I hope you have picked up, Newton's theory of gravity is a good one and is explained as a force propagated between two bodies. It is however not accurate enough to explain some of the newer observations of certain phenomena throughout the universe. Einstein's theory of General Relativity is consistent with all available observations and is explained as masses distorting space-time in their vicinity, and other particles moving in trajectories determined by the geometry of that space-time. Sounds like everything is pretty well understood then, unfortunately that is not quite the case as it has been recently understood that General Relativity was not quite as perfect as first thought. Unfortunately, the ultimate question of why atoms attract one another is still not understood. The goal is to combine gravity, electromagnetism and strong and weak nuclear forces into a single unified theory

The study of this unified theory is one of the most important areas of research in all of physics and is known as the field of Quantum Gravity. The aim of this field is an attempt to unify quantum mechanics (which explain the three fundamental forces of nature; electromagnetism, weak interaction, and strong interaction) with the general relativity which explains the forth (which is Gravity). Gravity is incidentally the weakest of all four of the fundamental forces by a massive amount. The electromagnetic force is roughly a million million million million million million million times greater than the gravitational force. It is actually so weak that we only notice it because it can act over such enormous distances and it is always an attractive (never repulsive) force.

So the next time you fall over

I bet you never thought that your simple act of tripping on your shoe lace or spilling your pint or getting soaked when it rained was quite so complicated. So the next time you do fall over and suffer the process of gravitation, why don't you give it a little more thought. If nothing else, it might at least distract your attention away from your embarrassment. If you are anything like me you will have become pretty well confused, pretty darn quickly, as the German rocket scientist Dr Wernher von Braun once said, "We can lick gravity, but sometimes the paperwork is overwhelming." And I couldn't agree more.

The importance of water

"In every glass of water we drink, some of the water has already passed through fishes, trees, bacteria, worms in the soil, and many other organisms, including people...Living systems cleanse water and make it fit, among other things, for human consumption."

Elliot A. Norse

If there is any single chemical substance that we criticize, curse and complain about more in the UK, probably the world, I can't think of it. Yet it is the epitome of irony that however much we hate water when it falls as rain, it remains the most essential substance for the survival of all known life in the universe (as far as we know).
Coke might taste better, but when you are really thirsty, nothing else will do, your body cries for water.

I am personally very protective of my water; at rugby training I used to have a large bottle with the words "DANGER! DIHYDROGEN MONOXIDE! DO NOT DRINK!" which I usually try and place near the ground-keeper's equipment, surprisingly quite a few people actually fell for it. Obviously not realising that dihydrogen monoxide is just an elaborate (and technically inaccurate) way of writing 'water'. The illusion was always short lived however, and was broken the instant any of my team mates saw me take a drink out of the bottle.

So what is water?
Mostly found in the oceans, water is the most abundant chemical substance on our planet and it contains some very unique properties which make it so important to life on earth. Water is both odourless and tasteless and although it may appear colourless, it actually has a very subtle blue hue, which is only visible when it is in very high quantities.

One molecule of water contains two atoms of hydrogen which is covalently bonded to one atom of oxygen, hence the formula H_2O. The molecule is a triangular shape with the slightly smaller hydrogen atoms attaching to the oxygen atom at an angle just greater than 90degrees from each other, exposing a large amount of the oxygen atom on the other side. This is very important as the other water molecules are attracted to the open part of the oxygen atom to form what is known as a hydrogen bond, between the water molecules. The hydrogen attraction is slightly less powerful a bond than those holding the oxygen to the hydrogen atoms but it is essential in holding the water molecules together, therefore creating water.

Water can be found in three different states, liquid, solid and gaseous. The melting point of water is 0 Celsius at which point it becomes a liquid, the boiling point of water is 99.974 Celsius at which it becomes vapour. Below 0 Celsius water is a solid. Unlike all other non-metallic substances, ice (solid state) is actually around 9% LESS dense than water (liquid state). If this were not the case, icebergs would not float and more importantly for many, ice cubes would sink to the bottom of your glass. This is caused by the way in which the ice freezes at the particular pressure that is present on the surface of our planet, which allows the density of the water to begin to decrease as the temperature drops. Water is at its most dense at 4C and least at 0C; ice at 0C is even less dense due to the way that the hexagonal ice crystal line up with each other. When the water is frozen the molecules of water are lined up against each other less efficiently due to the formation of the ice crystals which leave spaces and gaps which in turn reduces the density of the ice further. So simply put, the colder temperature affects the ice crystal ability to line up straight therefore, leaving spaces between the crystals and reducing the density of the ice and allowing it to float on top of the denser water. Ices ability to float is very important to fish underneath them and the mammals on top of them, if ice sank the world would be a very different place and life on earth would really struggle, at least in its current form.

Why is water so important?
Biologically speaking water is unique amongst all other substances on Earth as it is the single most important chemical substance for the existence of life. When we are looking for life on distant planets most of the time we are looking for water or at least evidence of the existence of water in the distant past. This is because as far as we know, water = life.

Water is also important as both a solvent and as an important part of many metabolic processes, without water chemical reactions would be unstable and no forms of life (no matter how basic) could function. This is clearly evident in the simple fact that all life forms are made of almost entirely water. A bacterial cell consists of 70% water, a human is between 60-70% water (dependant on how fat they are) the average plant is 90% water and the average jellyfish can be up 98% water.

One of the first things you learn about in Biology class in school, and possibly one of the reasons many students are scared away from biology, is the water cycle. So I am not going to go into it in too much detail, just in case there is still some of that left over resentment lingering. That said, to understand the importance of water you need to understand the water (hydrologic) cycle as it is vital to life on this blue planet of ours. Sun evaporates water from the oceans, lakes, plants and animals into the air, rains downs from clouds back onto the earth and oceans; runoff ends up back in the sea and starts over again. Easy...A+. Joking aside, the water cycle is extremely, extremely, extremely important to all life on earth and that's why it is one of the first things that everyone learns in biology class.

Importance to humans

Life on our planet does very well for itself *in* the water and when it made the big belly flop out of the water it still tends to congregate and flourish near areas with good sources of water. As I am sure you are well aware by now, this is no coincidence as life requires water more than anything else.

Humans are no exception; it may be possible for a person to survive a few weeks without food but lasting more than a few days without any water is very difficult. Although it varies from person to person and is dependent on conditions such as exercise, environment, humidity, the amount of water that should be consumed each day can be calculated.

Normal physiological activities, such as respiration, perspiration and urination cause the body to lose fluids, which must be replaced if the body is to maintain proper functions. It has been calculated that food contributes just under 1 litre of water and the metabolism of protein, fat, and carbohydrates another 0.3 litres. This means that on average about 2.5 litres of water for men and 1.5 litres of water for women should be drank each day, to restore lost fluids. It is important to note that this doesn't meant that you must drink 1.5 litres of straight water each day, beverages such as coffee, soft drinks and fruit juice will also do the trick.

How water shaped human evolution

Water is important to humans and as a species we have some unique water related adaptations. Our hairlessness for example, means that we sweat water from all over our body which can significantly cool us, stopping us from overheating. This allowed humans to become highly effective pursuit predators with much greater stamina than pretty much anything they were likely to be hunting, a trait echoed in marathon runners. However, it is the frozen form and the salty kind of water that has had the greatest impact on the path of human kind.

You may have heard about the ice age and about how the entire globe was smothered in glaciers of thick ice, sometimes kilometres thick. More specifically an ice age describes a geological period of long-term reduction in the average temperature of the Earth's surface and atmosphere, causing the expansion of continental ice sheets, polar ice sheets and alpine glaciers. Technically speaking we are actually *still* in an ice age, just a warm part of one. We still have a lot of frozen parts of the world, polar ice sheets, glaziers etc. In the entire 160 million years that the dinosaurs ruled the earth, they never saw anything as harsh as the winters that we still get on earth. However, conditions used to be MUCH worse and it was one of these major events that were vital for our species success.

Panama in Central America in nowhere near the "cradle of humanity" in south-western Africa where the human species is first thought to have evolved yet it's location may have been vital in the evolution of our species. The Isthmus of Panama has been credited with many milestones in Earth's history when it rose from the sea some 3 million years ago. It provided a land bridge for the migration of animals between the northern and southern America, forever changing the fauna of both continents. Vitally important for us however was the fact that it blocked a current that once flowed west from Africa to Asia and diverted it northward to strengthen the Gulf Stream. It is possible that this simple change in ocean currents may be behind the single most important event in our evolutionary history.

Simply put the development of the isthmus current may have made Earth more susceptible to a process known as the Tilt Cycle, and this created much more ice sheets, glaciers (ice ages basically) all over the Northern Hemisphere. The effect of all the ice was to be felt in Africa and it was to have a major impact on the evolution of the human genus. By causing the dramatic drying of the climate in Africa, the largely rainforest habitat that once dominated was replaced by grasslands. This forced our then tree dwelling ancestors to abandon their arboreal urges, start a new life on the savannah and eventually lead to our evolution. In other words, we would not exist if this little neck of land had not risen up across the ocean, thousands of miles to the west and locked up billions of tons of ice hundreds of miles to the north.

Conclusion
I don't know if I can write much more of a conclusion than, "water created humans" so I'm slightly stumped. I will just end with an interesting quotation by Ed Ford, who once said,

"Human beings were invented by water as a device for transporting itself from one place to another."

I'm sure he wasn't aware that humans almost were literally "invented" by water through the Isthmus current, though it does add an extra perspective to the comment.

Why dogs are stupid and cats are evil

"You call to a dog and a dog will break its neck to get to you. Dogs just want to please. Call to a cat and its attitude is, "What's in it for me?"

Lewis Grizzard

To be fair to mankind's most popular quadrupedal companions, the title of this chapter should probably read: "Why dog's *seem* stupid and cat's *seem* evil." Anthropomorphism is a pet peeve of mine and I would hate to be seen as a hypocrite.

I'm certain that we are all familiar with what I am referring to anyway as many dogs quite often act with a foolish, sometimes bordering on idiotic demeanour. Whilst cats often seem somewhat calculating, as if they are hiding an evil secret behind that ruthlessly cold stare.

Some examples of dog stupidity
There are endless examples of dog behaviour which can give the impression of stupidity and I'm sure you could think of more examples than I can. The seemingly "stupidest" dog that I personally know is my neighbour's dog, I'm sure they won't mind me using him as an example. His name is "Morton" and he is as daft as a brush, he also looks like a brush being that he is part spaniel, part poodle and part sheep.

Some of the most common traits that Morton has, which make him seem stupid include: barking at everyone and everything, running around the garden in circles and playing never ending games of fetch. The thing that gives the strongest impression of stupidity are his apparent mannerisms. The way he is always looking for attention, the way he always looks as if he is waiting to be told to do something and the way he always looks like he isn't quite sure what is going on.

Why are dogs like this?
First I must explain before you declare, "My dog doesn't act like that" I must explain that I use Morton as an example of exaggerated behaviour. Your dog might not act as stupid but I assure you they do share similar traits in their behaviour, if maybe not as obviously.

To understand why both animals behave in this way we only need to look at their living relatives. For the domestic dog (*Canis lupus familiaris*) that is the wolf (*Canis lupus*). Genetic analysis of the DNA of both domestic cats and dogs have in fact shown that both are actually in the same species as their wild counterparts, meaning in short that dogs are just a type of wolf. The dog's wild counterparts, the wolf, are where they get most of the characteristics of their behaviour. It is believed that dogs became domesticated by humans when groups of humans and packs of wolves (who often competed for food and water) found that it was mutually beneficial if they "befriended" each over. This obviously did not happen instantaneously and probably took many generations of increasing levels of tolerance, leading to trust between the groups. Wolves, although fearsome predators are not as dangerous as their popular image insists.

The most important attribute of the wolf is the fact that it is a pack animal. This means that much like humans they are used to living with other animals and have a complex level of social understanding. It is this social ability of the wolf that has allowed them to develop their high levels of intelligence and ability to work with others. In wolf society there is and "alpha male" who is dominant over the whole pack. Every other member is inferior and acts in an inferior (submissive) way to the dominant leader. The pack works sort of like a cast system; there is a scale of superiority with some members being higher and some being lower. Each member knows their place and will generally not challenge it outside of extenuating circumstances. It is actually this social system which accounts for the trainability, playfulness and general abilities to bond with humans. It is also the reason that sometimes dogs seem to be acting stupid.

Back to Morton

Morton isn't a stupid dog. The reason he acts the way that he does is actually due to him simply behaving like his wolf relatives. Like all pet dogs, Morton is at the bottom of the social order amongst his pack (which he considers humans to be part of.) If you have ever seen a dog roll on his back, this is a perfect example of this. The dog is acting submissively and presenting you with his most vulnerable body part as if to say "I completely trust you and I give myself over to your superiority." The playfulness of dogs and the hyper-activeness is also explained by the dogs need to try and fit in with the pack and keep his pack-leaders happy. They always jump up and run towards you because they are showing you that they want you to like them, in a sense 'be their friends. Dogs have very similar natural ideals towards many important human attributes including friendship and love, due to their pack mentality. This both explains both why they always want your attention as well as why dogs will (usually) do what they are told.

So put simply, Morton barks all of the time because he is eager to protect and defend his pack. He runs round in circles because of he is a pack animal that enjoys social interactions which playfulness is a very important part of for bonding and establishment of dominance in some situations. He always wants to play fetch because he is eager to please a dominant member of the pack, and he thinks that fetching for you is pleasing you as well as the aforementioned importance of play in pack life. One of the main things that make dogs seem stupid is the expression they often have on their faces, that sort of cute, excited and sometimes slightly idiotic look they give you. This expression too can be explained by pack mentality and has evolved to help display that the dog is content and not aggressive. It is an expression which means "everything is ok, you are the boss, and I don't mind that" and helps to keep the peace and reduce disputes within the pack. Just to us, it makes often makes them just look dopey. When dogs were domesticated the tamest individuals were chosen, resulting in this cute look being exaggerated, as long as the submissive behaviour.

I could go into greater depth and explain all the various behaviours of your pet dog using wolf behaviour, as all dog behaviour is just wolf behaviour with a PG certificate.

"Some people say that cats are sneaky, evil, and cruel. True, and they have many other fine qualities as well."

Missy Dizick

If dog's apparent stupidity can be explained by their pack mentality and is actually little more than behavioural misunderstandings on our behalf. What about cats? Well, with the dog's arch nemesis things are a little subtler. The Feline facade is often of a more intelligent, cunning and altogether, at least in my opinion "evil" animal, but is this really the case? Of course the domestic cat is not evil it is such a popular pet due to the fact that they have the ability to be both affectionate and loving towards humans.

Some examples of cat evilness
To be honest I think that it's the overall attitude of the cat which makes it seem slightly evil but there are also several trait and behaviours of

domestic cats which seem to back this attitude up. Behaviours such as always watching what you are doing all the time, the common trait of approaching the one visitor who doesn't like cats, toying with small animals it has caught and bringing home dead (or almost dead) prey and leaving it for you to discover later.

The most "evil" cat to which I have ever had the pleasure to briefly meet was my Grandpa's cat (I don't even know its name). This cat lived in Spain and was almost never around; it would spend sometimes days away for the home, down in a small valley a few hundred yards away. Every few days the cat would arrive home and would often bring with him evidence of what he had been up to all that time. Often he would leave on the porch the sometimes half-eaten but usually just dead remains of his latest kill. This would range from mice and rats, to lizards and snakes and even the occasional rabbit. This cat was not a house cat, although he did start out that way. The house still acted as his "den" or something similar, a place where he could seek shelter, safety and security if needed. The cat hardly ever showed any affection, or even paid any attention to the houses human inhabitants. He had his own live, and as far as he was concerned his supposed "owners" had very little to do with that except for occasionally supplying water and food.

Why are cats like this?
 The domestic cat (*Felis silvestris catus)* is actually a type of wildcat (*Felis silvestris)* and are merely a domestic version of this species. A 2007 study discovered that all of today's domestic cats are the relatives of as few as five African Wildcats (*Felis silvestris lybica*) from around 8000BC. It is believed that these animals actually introduced themselves into human habitation probably due to the large number of mice and rats attracted to human food-stores. Both humans and cats were to benefit as the humans now had a highly effective organic pest control system which eventually became the close "friendship" which we have today.

As with dogs, all of the behaviours which we see displayed in domestic cats are a throwback to their wild roots. Unlike the dog, cats are not pack animals, however they have evolved behaviours which allow them to be highly tolerant of other cats and even live with each other in relative peace. Most of the behaviours which you see in cats such as purring, bumping you with its head, rolling on their backs, hissing and tail flicking have all evolved as a means of communicating the individual's mood or displaying affection. Both of which are very important if you are going to have to try and get along with other individuals.

To be fair to cats they are not evil, in fact compared to most wild cat species they are nothing but pussycats. The apparent evil attitude of cats, again, is nothing but a miscommunication. The reason cats will often watch us and even, to quote Vicky Pollard from Little Britain, seem to "give us evils" has nothing to do with any sort of malicious intention but is actually a cry for attention. The reason cats very often walk straight up to the person who doesn't like cats or will look for attention when you are on the phone or watching TV, is not a display of the cat's sick sense of humour. In reality the cat is responding to your body language. In "cat society" when a cat threatens another it will stare intensely and move gradually closer to the cat until one cat either runs away or a fight ensues. Someone who doesn't like cats or is ignoring the cats will be doing the opposite, looking away from the cat and generally acting in a non-aggressive and therefore cat-friendly manner. My advice to people who don't like cats, stare out the cat before approaching it slowly, nine out of ten times the cat will run away.

The most "evil" behaviours which cats display are the manner by which they continue to stalk and kill small animals even when they are well fed. Again, this is not the cats fault and as with everything in life is a result of its genes. Cats are carnivores, and they are predators, because of this they have an innate instinct to hunt and to kill what is of course their natural prey. This instinctual behaviour can also be seen in all cat play, when they are merely practising hunting and pouncing and killing techniques which they intend to take outside as soon as you let them (or whenever they feel like, if you have a cat-flap.) They are in fact so good at killing that the RSPB (Royal Society for the Protection of Birds) has made a plea to cat owners, not to let their cats out at night because they needlessly kill so many garden birds.

Cats do kill small animals due to their innate behaviour, however some scientist have suggested that the domestic cat might in a sense be losing some of this behaviour. Many cats seem to no longer make the connection between the "killing" and "eating" of prey. This explains why cats will often bring prey back to their home but will not eat it, as they now receive food from us they no longer link hunting of small animals to finding a meal, and therefore fail to make the final connection. Another explanation is that they are so well fed that they simply don't need to eat the food they catch. This explanation however doesn't quite tie up with stereotypical predatory behaviour whereby the predator will eat all they can as they never know when the next meal will be.

The act of playing with their "food" although frowned upon by polite society it is a generally common behaviour in carnivore society, particularly feline society and although it does seem extremely malice, it is an important one at that. Nearly dead prey is often brought back for the young to practise their killing techniques on when they are not quite ready for "fully mobile" prey. Adult cats do the same for themselves basically because cats enjoy it.

Big babies
Both domestic cats and dogs like to play so much that it has been suggested that both have been selectively bred in a manner which promote childlike behaviour. Both cats and dogs act very much like puppies and kittens for most of their lives, probably due to humans (either intentionally or possible by coincidence) have bred them to be that way. Childlike pets are always going to be more affectionate, less aggressive and more playful. It also explains why both domestic cats and dogs remain submissive their entire lives, because they are really just big babies. Dogs seem stupid, cats seem evil but both really just want to get along with their larger, less hairy contemporise.

The dog has undoubtedly been the more successful in his relationship with man and many have suggested that without the aid of the dog man would not be in the prime position he is today and had a powerful influence on our development as a successful species. Dogs have helped us catch food, herd animals and protect us from dangers for thousands of years and we have been co-evolving in a truly unique way. Behavioural studies have shown that there are more similarities between the way we think and the way a dog thinks that between us and a chimp. We have thousands of years of mutual reliance to thank for that. The cat on the other hand has been the more successful at assimilating itself into our lives without having very much of an impact at all and is its greatest strength.

So in conclusion dogs are certainly not dumb and cats aren't evil either...just good pets.

Does Astrology work?

"I don't believe in astrology; I'm a Sagittarius and we're sceptical"

Arthur C. Clarke

Predicting the future has always been one of the primary focuses of the human race, that's why things like astrology remain so very popular, even in these modern and supposed "enlightened" times. Astrology as I am sure you are aware, is the claim that events on Earth are related to, and affected by the apparent positions and motions of celestial bodies from across the vast expanses of the universe.

This is actually a pretty easily scientifically testable supposition, or at least it would be if any astrologist ever allowed themselves to be tested. The problem is (and it's probably wise as far as they are concerned) that they never make any specific or definite predictions that could be accurately tested. Instead the predictions which they make are always so vague that they could be interpreted to mean almost anything. The statements that astrologers are designed so that the prediction can be read and interpreted in many different ways, which means it is near impossible to prove to disprove that they mean anything specific, and therefore can never be proven as incorrect. For example, making statements such as "you will encounter a financially rewarding opportunity" or "relationships may become turbulent" can mean a number of things and can't be proven as inaccurate. As you can imagine, both of these statements could be interpreted in a thousand different ways, dependant only on how imaginative you are, and could never be "pinned" down to target any one particular result.

The "science" issue
The real reason that astrology does not lend itself to scientific understanding is not the lack of scientific evidence but more that fact that it is not consistent with other areas of science which are well understood. For example, even the ancient scientific research of Copernicus and Galileo (which proved that the planets actually revolved around the sun and not that the Universe revolved around the Earth) make astrology become literally impossible by its very definition. Why would the positions of other planets as they happen to be seen from

Earth have any sort of relationship to the atoms moving around on a tiny insignificant planet such as our own, but this is what astrology would have us believe?

How "real" science compares
Some of the most complex and technical science around actually threatens to almost give astrology an outside chance of sounding plausible, just by its very complexity. There is actually not much more experimental evidence around for some of the world's most important scientific theories that there are for astrology. The difference is that they are supported by and tie up with other theories which have survived large amounts of scientific testing for many years. For example, well known laws of physics such as that of gravity and many other similar and related laws lead to something called "scientific determinism". This states that if we were to know the positions and velocities of all the particles in the Universe at one time then it would be technically possible to predict the state of the universe at all other times in history. Basically scientific determinism states that it is therefore possible for us to accurately predict the future, which would mean that astrology would become pretty unnecessary. Of course, all though possible in principle when looked at in practise even the most basic and simplistic calculations that would need to be made, do not hold out for more than one or two individual particles for only a few fractions of a second. To make things even more complex, there is a property known as "Chaos Theory" which means that even a tiny change in position or velocity at one time can lead to completely different behaviour at another time. I'm sure anyone who is familiar with "Jurassic Park" will remember this explanation. Basically it states that sequences of events cannot be repeated exactly the same more than once, there is always going to be something that happens which will change the effects of something apparently completely unrelated, resulting in different result. The old example always given is that a butterfly flapping its wings in Hong Kong will trigger a tsunami on the other side of the world. It is said that the flapping will cause a series of events which are linked together in a huge chain reaction, building all the way up to a natural disaster on the other side of the world. However, chaos theory also states that this could never happen more than once, the next time the butterfly flapped its wings it would trigger a different event and because it is a different time, all of the other stages in the chain reaction would also be different and not conclude in the same outcome. This is why the daily weather forecast can be so hard to predict, there are just so many factors which are never remain

constant and all affect each other in different ways under different conditions.

Therefore, although all things being equal the laws of Quantum electro-dynamics in principle make it not impossible predict all the things which astrologers claim to be doing, it is not possible in practise as there are just far too many factors and variables and uncertainties to consider. In fact, "the uncertainly principle" states that the closer we actually get to calculating all the data, the further we will actually be from accurately calculating it. But enough of the theoretical physics, I didn't even study Physics at school so I'll probably start to confuse myself if I go any further. It does however do a pretty effective job of making astrology look pretty silly, which is why I felt it necessary to mention.

So what do astrologers claim to be doing then?
The truth is that astrologers have (rather unsurprisingly) shown themselves to continuously be unable to explain how astrology is actually claimed to work. Even when not asked to directly demonstrate the process being carried out but only explain what they claim to be doing, they are still not able to give any explanation. So basically, they won't even tell us how it is meant to work, suspicious much? Astrological organisations now actually state that they no longer claim a direct relationship between the planets and events on happening on earth. They must have realised it was just a tad too far-fetched. Instead astrology has now become even less of an exact science as different astrologers claim different explanations towards what they are actually doing. In most cases the mechanism has become much less mechanistic and much more based around divination, psychic predictions and other similarly related twaddle. Alternatively, there also remains some astrologers who believe that their techniques are indeed pliant to the scientific method, stating that it has even been validated by statistical analysis. The reality is that the scientific community has continuously demonstrated astrology failing to prove its effectiveness. Studies have shown that the accuracy of the prediction made using astrology is actually no better than would be expected by chance. People might find themselves believing the astrological reading they get by (often sub-consciously) exaggerating positive hits and ignoring anything that doesn't quite fit. The truth is that the only real correlation that seems to be able to be demonstrated between science and astrology is that the more detailed the scientific study used, the weaker the performance of the astrology.

Conclusion

With scientific testing a side it is still very easy to test the accuracy of astrology. Read the daily prediction of all the astrologers that you can find. If there was any actual method to astrology, then then every astrologer would get the same results each time they make a prediction. Therefore, should if they don't all match each other then they can't be all true. Still not convinced, well consider this. People always think that their astrological prediction is a custom reading. However, the reality would be that this would mean there are only ever twelve different type of thing happening to everybody, each day, in the entire world. And that every day, each person will have the same things happening as every other with the same star sign. Not only that, but astrology implies that they also all have the same personality, temperaments etc. Having an identical twin myself, you can imagine I'm personally not wholly convinced.

The Science of Guinness

"Alcohol, the cause of and solution to all of life's problems"

Homer Jay Simpson

Guinness is a mysterious drink. The lifeblood of the burly, Irish gent and possibly Ireland's most famous and celebrated export. There are also many highly celebrated mysteries contained within the "good stuff" which separate it from "normal" beers. Other than why 'does it taste so unique', many also wonder why the head is that lovely creamy white, colour when the beer is so deathly dark, how does it keep its head and why do the bubbles sink?

So many dark mysteries...
To begin with, Guinness as opposed to other similar dark beers (Toohey's Old, Newcastle Brown Ale etc) uses nitrogen, instead of just carbon dioxide, to get its bubbles. Nitrogen is used as that is the style of an Irish Stout, and it is what gives the beer a smoother taste. The smoother taste is achieved as in normal beers when the head is produced by the solution of CO_2 and water which creates an acid (carbonic acid). The rich creamy head in a pint of Guinness is achieved because nitrogen bubbles are much smaller than CO_2 bubbles. The head also lasts longer as nitrogen bubbles do not continuously grow as the CO_2 comes out of solution in the beer. As a result, the nitrogen remains in the beer for a longer time, causing a much longer lasting creamy head. So long lasting indeed that it usually remains until long after the pint has been finished, and a fresh pint ordered.

Another mystery of Guinness is why is the head is white whilst the rest of the pint remains so deathly black? The reason for this is due to the way the light travels through the head, the liquid never actually changes colour at all. As explained earlier, the bubbles in a Guinness are very small. As a result, the foamy head contains an extremely large number of bubbles and not much beer. Light entering the foam is therefore rapidly scattered in different directions by multiple encounters with the bubbles. Some of the light finds its way to the outside, and because all wavelengths are affected in the same way, we see the foam as white. This process is known as Mie Scattering, and is a similar process to what makes clouds, milk, white paint and polar bears appear white when they are all technically transparent. The same cannot be said of the actual beer of course, which light has more difficulty escaping from, so we see as dark brown to black colour depending how bright a light we shine on it.

One of the most popular Guinness related bar conversations is in relation to the famous Guinness Cascade. This is when the bubbles in the liquid appear to move downwards, apparently defying the laws of gravity. The cascade is in fact a result of fluid dynamics, a process known as hydrodynamic drag. Bubbles that touch the sides of the glass are slowed as they travel upwards, yet the bubbles in the centre move to the surface at normal speed. This creates a rising column of bubbles in the centre of the pint glass which creates a rising current by the entrainment of the surrounding fluid. This current makes the beer near the outside of the glass flow downwards. The effect occurs in all liquids but is just more noticeable in a pint of Guinness due to the contrast between the dark liquid and light bubbles.

The final mystery about a pint of Guinness is how it is able to maintain its head throughout its entire drinking life. Even if you directly and deliberately remove the head, a new one will quickly form from the dark liquid below. It's a trick that no other beer, ale or stout can carry out as efficiently as Guinness and it's again due to its unique fluid properties. What is keeping the head in place known as Isostatic Equilibrium and it explains how two layers of different densities and thicknesses interact between each other and stay in balance. When the mass of one layer is altered the other layer will alter itself "automatically" to re-dress the balance through isostasy. So as the head is removed, the fluid from beneath rushes up to replace the lost head. It is similar to what we see in a dynamic natural system such as plate tectonics, with the formation of mountains and sea level changes, and it is actually from geology that the terms above are taken from.

Conclusion
So that was "The Science of Guinness", but what would the chapter be without a quick reference to the old Guinness slogan "Guinness is good for you"? Although I can't condone a beer based diet, there have been studied that have suggested that Guinness may indeed be good for you. Some studies have found that it may contain antioxidants similar to those in fruit, which may slow down the build-up of cholesterol on arteries, which my help your heart health.

So as well as being great for teaching you about science, Guinness may also be helping your health slightly too. Guinness is good for you!

The Science of Fire

"Man is the only creature that dares to light a fire and live with it. The reason? Because he alone has learned to put it out."

Henry Jackson Vandyke, Jr.

One of the most significant turning points in the cultural evolution of early humans was the ability to control fire. Working out how to successfully manipulate fire allowed for the rapid expansion of the recent human ancestor *Homo erectus,* first and foremost by allowing them to gain more nutrition from each meal. By cooking food, it allowed many of the previously inedible parts of plants and animals to be made edible. Also it allowed the access to higher levels of important macronutrients such as proteins and carbohydrates to be gained, from the now more easily digestible food. Early humans would obviously be unaware of the benefits of cooking, so would not learn to cook for that reason. It is more likely that humans would collected burnt up carcasses of animal's killed in forest fires (as they still do today) and discovered it was easier to eat. Fire had the added benefit of improving the survival capabilities of these populations by providing a readily available source of heat for warmth, light to allow for night-time activities, protection from predators and later on in human evolution, as a powerful tool for landscape management and hunting.

Man make fire...?
The ability to simply control fire probably came long before the more technically challenging ability to create fire emerged. It is likely that early hominids exploited natural sources of fire such as forest fires and lightning strikes. They would have possibly kept burning embers with them at all times and became skilled in keeping the fire 'alive' by feeding and generally looking after the burning ember as if it were a living thing.

Although at the time it would be unlikely that they would have been aware of the importance that fire would ultimately have on future generations of humans, it is undoubted that fire would have been considered precious to early humans. Many early cultures and societies viewed fire as some kind of divine gift and centred much of their rituals and religious practises around the fire. Today fire is just as important to us as it ever has been. Many of its current uses remain primitive; some have been adapted and expanded to suite modern demands. One thing remains constant however, just like during prehistoric times most people have no idea what fire actually is.

So what is fire?

What most people will think of when they visualise a fire is a flame, as the flame is the visible part of a fire. However, the true definition of a fire is that it is a chemical reaction which results in a combustion chain reaction that keeps the fire burning away. Fires occur when a flammable material is supplied with a sufficient supply of oxygen (or similar oxidizer) and a source of fuel. Exactly how likely a fire is to occur depends largely on the flammability of the material, which is defined as the ease of which the substance will ignite.

All materials can be tested and rated for flammability from 100% or A1 (which is not flammable) to 0% or B3 (which is easily ignited) for example, wood is classified as B2, or normally combustible.

A fire can't exist without a source of oxygen or some sort of oxidising agent and doesn't necessarily have to contain oxygen. To form an affective oxidiser a compound must either readily transfers oxygen atoms or gain electrons in a redox reaction (a reaction in which atoms have their oxidative state changed). Fuel is the case of a fire is defined as a material which releases its energy through the chemical reaction of combustion by reacting with substances around them, namely oxygen and heat. Obviously all of the components of the fire are highly variable and will alter the type and temperature of fire which is created.

Now you understand what it takes to make a fire I'll go back to the flame, which is probably the most important and certainly the most easily identifiable part of a fire. I would also probably not be wrong if I were to say that the majority of people are completely unaware of what a flame actually is, even though they might see one every day.

Well technically speaking a flame is an exothermic and self-sustaining part of an oxidizing chemical reaction which produces heat energy in the form of glowing hot matter and plasma. This definition should be easily understood by the descriptions that I have given, that is except the word plasma. A flame contains within it a state of matter which is known as plasma, which means that it is partially ionized gas. This means in that a proportion of the electrons, instead of being bound to atoms or molecules are free. This means like unlike solids, liquids or gases, plasma has neither a finite shape nor set volume, making it a unique and distinct form of matter that many people are unaware it even exists. When you are taught about the solids, liquids and gases at school, they miss out plasma (and several other even more complex states of matter) to avoid confusion.

The colour of a flame is also an important factor in any fire and can tell us a great deal about that fire from its temperature, the source of fuel and even where the fire has come from. For example, many different chemicals will burn with different colours and this will also be the case for the fuels in which the chemicals exist. Methane gas (CH4) is a good source of fuel and it burns with a mostly blue flame; coal (a rock containing mainly C and H) is also a useful fuel but has a more orange flame; and we all know about magnesium (Mg) which burns with a glowing white flame. The colour is significant here as a red flame such as is sometimes visible in a forest fire indicates the lowest temperature fire (from 525C-1000C). An orange flame, which can be seem in hotter wood and coal fires, a lit cigarette or a match indicate a hotter fire (from 1100C-1200C). A blue flame such as seen in a Bunsen burner or blow torch an even greater heat (around 1300C); and finally a white flame is the hottest fire of all (from 1400C-1600C) and the more brilliant the flame the hotter. Magnesium for example, although you are unlikely to encounter in most fires, burns at 2200C but only needs 425C to ignite it.

Fire, our friend
Although everyone is aware how potentially dangerous fire can be, ultimately fire has been humanities friend ever since we picked up a burning stick and realised that it was hot. We rely on fire for so many things that it would be ridiculous to try and list them all, so I'm not going to attempt to. However, whether using fire to cook our dinner or for creating a romantic candle-lit atmosphere in which to enjoy it are two of the best. Fire is more than just another exothermic reaction. Fire is our heat from the cold, our light in the dark and our protection from danger. It's been with us from the start, our saviour through hardship and our partner in crime.

Is Time Travel possible?

"If we could travel into the past, it's mind-boggling what would be possible... I have no idea whether it's possible, but it's certainly worth exploring."

Carl Sagan

Is time travel possible? Could a significantly advanced civilization go back and change the past? If we are to take Einstein's Theory of General Relativity seriously, we must keep in mind the idea that space-time may tie itself in a knot and information might get lost in the folds. The serious study of time travel is not something that is widely studied by many scientists. This is mainly as a result of the potential outcry at the massive wastage of public money, but also due to the apparent terrible dangers of studying such a field. After all, what is to stop the wrong person going back in time and changing the future? Someone with a time-machine could quite easily go back, change history and rule the world! Maybe it has happened, how would you know. The existence of fossilised human footprints might be a giveaway.

There are indeed only a very few physicists whom are full-hardy enough to study the field of time-travel. These individuals even find it necessary to use their own special kind of code for time-travel research, as to avoid the ridicule of their peers. The well-known Astrophysicist, Stephen Hawking spent much of the 1990s trying to show that there was a law of physics that prevented time travel from being possible, which he called "the chronology protection conjecture." Though after many years of hard work he failed and now concedes that time travel is perhaps possibly possible. Kip Thorne, a close friend and associate of Stephen Hawking, is also currently one of the only serious scientists with enough courage to openly discuss the reality of the possibility of time-travel. Stephen Hawking mentions in his book "The Universe in a Nutshell," that he and Kip Thorne often make bets against each other on various topics of personal research. However, in regards to Kip's research into the possibility of time travel, he and Hawking are now both supporting the same side. They both allow that time travel of macro-molecules is possible, although highly impractical. By macro, I am referring to objects which are large in molecular terms. A spaceship would be a macro-molecule; an atom is not.

(Time-travel) back to the start...

The basis for all study in the field of time-travel is Albert Einstein's famous General Theory of Relativity. A detailed study of Einstein equation of general relativity explains how space is dynamic, as well as describing how it is specially shaped and curved. In general relativity, someone's personal time (as measured by their watch) would always increase, as it does in Newtonian theory or the flat space-time of special relativity. However, there is in reality, the possibility that space-time could be warped so much so, that someone could leave off in a spaceship and arrive back before they had even set out. One way in which this would be possible, would be as if there existed such a thing as "wormholes." Wormholes are tubes of space-time that link together different regions of space and time. The basic idea being that you could drive your spaceship into the mouth of a wormhole, and exit at the other end of it, finding yourselves in a different place and a different time.

Wormholes if they exist (which they probably do) would be the solution to the speed limit problem in space travel. The reality of space travel is that it would take tens of thousands of years to cross the galaxy, if travelling in a spaceship travelling at less than the speed of light, which general relativity demands. However, if travelling through a wormhole you could travel to the other side of the galaxy and back home again, easily in time for dinner. If wormholes exist, it would even be possible to use them to get back home to a time BEFORE you had even set out! This unfortunately creates a serious problem for time travellers known as the "Grandfather Paradox." What happens if you travel far enough back in time that you are able to kill your own grandfather, and prevent the birth of your father? Will this mean that you will suddenly cease to exist? I am not going to go into discussing free will, or whether or not you have the free-will to do whatever you want when you travel back in time. I only aim to discuss whether the laws of physics allow space-time to be so warped that a macroscopic body (such as a spaceship/person) can return to its own past.

According to Einstein's theory of general relativity, a spaceship must travel at less than the local speed of light, and follows what is known as a "time-like path" through space-time. This means it is possible to formulate the question in technical terms "does space-time emit time-like curves, which are closed?" i.e. ones which returns to their own starting point, over and over, again and again. These paths are known as "time-loops" and it is these which can be followed in order to travel through time.

The important question that needs to be asked is whether a significantly advanced civilisation could create a time-machine. To do this they would need to be able to develop a machine that was capable of creating an area in space in which concentrations of time-loops could be made to appear in a single, focused and finite region of space. Time travel is therefore technically possible according to the laws of physics and has recently been shown that it is very likely to be constantly occurring on a microscopic scale. It is very unlikely however that it is ever going to be able to work on a macro-molecular scale. It becomes extremely unlikely therefore that a time machine is ever going to be made. The actual physical reasons that time travel is not likely to be able to work on a macroscopic level goes far beyond my ability to explain to you here. The main obstacle in the way is the need to assemble astronomical amounts of energy, roughly equivalent to the mass of a stellar black hole. What's more, Kip Thorne has also showed that keeping a wormhole stable would require something called negative energy – an exotic phenomenon in which quantum fluxuations render the energy density in a region of space less than zero. It is a technical possibility, but it would rely on far too many outstanding factors going exactly to plan and never going wrong ever. The chances of that occurring are pretty much nil; certainly it is only really conceivable for a civilisation significantly more advanced than ours.

Conclusion
Some people speculated that when the Large Hadron Collider was turned on at CERN in Switzerland that time travellers might suddenly hop out of somewhere. I'm not sure why, I think people got slightly carried away and maybe didn't realise that particle accelerators are not exactly a new idea. However, as far as we are aware this did not happen. If it had it might have been the single most important event in all history, so it's a shame really.

My final conclusion is that time travel is possible... well it is essentially feasible. Ok it's not going to happen! Stephen Hawking has calculated the chances of time travel being possible at being somewhere in the region of 1 in 10 with a trillion, trillion, trillion, trillion, trillion zeros after it. Not being a gambling man I would probably consider that a losing bet, but with odds like that imagine the pay out! Any sensible betting shop would probably not offer any odds to someone willing to put any serious money on time travel, odds on they might be from the future.

Why is human childbirth so painful?

"I would very much rather stand three times at the front of battle than bare one child"

MEDEA, from the ancient Greek tragedy of the same name

Now I don't know if Medea ever did get the chance to taste battle but the wife of the legendary Jason (of Argonauts fame) did have two children, so we can't say she doesn't know what she is talking about to some degree.

It is often claimed that women have a higher pain threshold than men. I have heard this said to my many a time, usually when I'm sick to some degree or suffering from another rugby injury. It is always women that say it with a supercilious smirk, and I relish in bursting their bubbles of feminist empowerment. Sorry ladies, but there is no research to support your favourite statement, in fact all the research shows that men have a much, much higher pain threshold than women. It's not even close... bummer eh? Oh dear, I've dug myself another hole. Anyway, this then makes it even worst that women have to put up with the excruciating agony of childbirth. No other animal suffers as much difficulty or tolerates as much pain as humans do when giving birth. No other animal seeks out or requires the assistance of other individuals when giving birth; most species actually look to be alone. Confident in their own safety during birth or egg-laying, they seek isolation to protect their vulnerable young from predation.

Human mothers on the other hand are extremely vulnerable during childbirth and until fairly recently death during labour was constantly in the back of every mother's mind, and for good reason. Prior to the modern advances of the 20th centaury, the mortality rates of human mothers during childbirth were much higher than that of any other mammal (except one, which I will reveal later)

This is a hard concept for many of us to grasp, we always place ourselves up on a pedestal as the pinnacle of evolutionary triumph (as if there were such a title.) The only animals to have mastered so many things that we see as great and unique accomplishments. From evolving complex language to mastering fire, fashioning tools to taming wild animals, splitting the atom to computer technology and everything in between. We have managed very well for ourselves all thanks to one key evolutionary advantage. The human brain, and it is the brain that makes childbirth so very difficult.

Why blame the brain?
Can't blame it on the sunshine, can't blame it on the moon light ... well you all know how it goes, and you certainly know that "good times" and "boogie" aren't in a mother mind while she is trying to birth a melon-headed infant. And the reason for this is all down to the brain, more specifically the SIZE of the new-born human brain when compared to the space it has to be squeezed out of.

 Well that's not entirely accurate; to blame the brain is sort of like arriving at a plane crash a blaming the pilot. Sure the pilot was at the centre of the plane crash, but you can be sure there were a lot of extenuating circumstances which lead up to the pilot bringing the plane down. Comparing child birth to a plane crash, possibly not my wisest of analogies but an affective one nonetheless.

 Anyway, back to my original point. To blame the size of our brain isn't fair as it is really down to a combination of factors that makes child birth more difficult in humans than any other species. Except maybe hyenas, which I mentioned earlier and will briefly discuss later to make the ladies feel better.

During hominid (human ancestor) evolution there was a point when our behaviour changed markedly. We moved down from the trees on a more regular basis, and as we spent less time in trees our walking stance became less adapted for tree living and batter at walking. As these hominids became more proficient at living on the forest floor and moved onto the savannah they became more upright in stance, which brought about secondary adaptations. No longer having the ability to run and hide up a tree and clearly not having the speed or strength to avoid becoming something's dinner, our ancestors took the geeky route. They developed their cranial capacity in an attempt to outwit potential predators and not become their dinner.

Unfortunately for our mothers and daughters, two-leggedness and large-headedness are not compatible. The rate of brain size growth over the last 2.5 million years has been 3 fold, whereas the size in the pelvic girdle hasn't kept pace, hardly increasing in size at all.

Unlike in all other quadrupeds, humans have a unique shaped pelvic girdle due to our upright stance. It is somewhat open, flattened from front to back and acts like a "bowl" which contains and supports all of the internal organs. If it wasn't for the shape of the pelvic girdle and the extremely strong anal muscles, there would be very little to stop a person's insides, from falling outside of their bottom. This is an issue that no other animal faces, and also one caused solely due to our upright stance.

It has a serious consequence for childbirth too. When a child is born it obviously has to pass through the pelvis. A newborn has many adaptations to make this process easier which I am not going to discuss. One adaptation that could exist, but doesn't and would giving birth easier and safer would be if the pubic ring of bone was larger, as the baby has to squeeze out of this just before it is born. This is the smallest hole the baby has to fit through and is the most dangerous part of the entire birth for baby and mother. Why can't it be bigger, just a little perhaps? For the same reason the rest of the pelvis is the shape it is, the hole just can't get any larger or it would be unable to support the internal organs affectively. So it is essentially not our large brains but our upright stance that make child birth so difficult. Walking upright has given us the advantage of a large brain but the disadvantages of a painful and dangerous childbirth.

In evolution everything is in balance, everything is give and take. If you want to come out of the trees, you will be more vulnerable to predators; if you want to walk upright, your internal organs will have to moved; and if you want a big brain, child birth is going to be hard work. That's how evolution works. You can't make something out of nothing, you make do with what you have and go with it.

Conclusion

As promised, I would end the chapter by trying to make the ladies feel a little better about the sacrifice they are making. You really don't have it all that bad; I mean you could be a young female hyena. The Spotted Hyena is an example of both painful childbirth and that 'make do with what you have and go with it' burden of evolution.

Female hyenas are exposed to very high levels of androgens such as testosterone during the later stages of pregnancy which is meant to promote aggressive and dominant behaviour, in this female dominated society. However, it has the added effect of hyenas developing very unusual genitalia, with the females having an enormously enlarged clitoris with more than a passing resemblance of male genitalia. Apart from making the women look and act rather butch, this isn't a problem right? Well no, the problem starts when the females have to try and reproduce. Because these 7inch pseudo-penis' are actually clitorises, they have to find real penises to reproduce with. I don't know how they do it, but I can imagine it's tricky. That's not where the real problem lies though. As the pseudo-penis is actually female genitals so once the hyena goes into labour she has to give birth through it. As you can imagine labour is often very lengthy, the organ become quite painfully stretched and often is torn to pieces. Up to 60% of stillborn have been recorded and 10% deaths amongst mothers. A much higher death rate than you would expect from a supposed evolutionary stable strategy. It is possible that because hyenas live in large social groups that are female dominated, that the impact of the high death rate is somewhat "buffeted" and its affect not felt as hard.

I have never seen a child be born, I have a reputation for saying the wrong thing at the wrong time so I probably wouldn't be very welcome. That said I would certainly find it comforting if someone said while I was going through childbirth, "It looks *pretty* sore, but don't worry it's not a painful as a spotted hyena's labour!" ... Am I wrong?

How do Whales become beached?

"Toward dawn we shared with you
your hour of desolation,
the huge lingering passion
of your unearthly outcry."

from 'The Wellfleet Whale' by Stanley Kunitz

I'm going to give myself the benefit of the doubt and assume that anyone that takes the time to read this book is generally the type of person who will find topics in which I find interesting, also interesting. More than likely I am not completely correct and I am sure there are one or two chapters that you might have thought about skipping and that's perfectly fine. I don't think I am wrong in saying that most people do however like Whales and Dolphins (collectively known as Cetaceans) so I think they are always a good and safe topic of discussion.

During the week that I was writing this chapter, a member of the UK's only resident pod of Orca was washed up dead on The Isle of Tiree in spring 2016. The female Orca named Lulu was identified by her unique scars, saddle path and eye patch. The Hebrides Whale and Dolphin Trust (HWDS) who have been monitoring the population, hadn't seen Lulu since July 2014. There are now just 3 females left in this fragile pod, which according to Orca expert, Dr Andy Foote, is too late to be saved from extinction.

The pod has not produced offspring in several generations, and there are now believed to be only 8 individual remaining. Dr Andy Foote believes the main culprit for the diminishing Orca population off the coast of Scotland is human based condiments. A study into the cause of death of Lulu confirmed the death was due to entanglement.

"There were deep, granulating wounds ... consistent with a rope wrapping around the tail and trailing behind the animal, probably still attached to something at the other end. This would have made normal swimming very difficult, and we suspect the animal had been entangled for several days. She hadn't fed recently but had swallowed a large amount of seawater, most likely as she eventually succumbed to the entanglement and drowned."

With news of a further case of whale entanglement in the same week, this time a Humpback Whale, in Loch Eribol, Durness, it is ever more obvious how at risk whale populations can be.

So what causes whales to become stranded? Are humans to blame for all of the whales washed up on to beaches?

What causes beaching?
As are constantly reminded in nature documentaries, cetaceans are extremely intelligent animals with highly advanced senses of direction and navigation skills. Something pretty serious surely needs to have gone wrong to lead a whale or dolphin or even an entire pod, to become stranded.

Unlike in the case, the type of standings which usually causes the greatest media attention and human sympathy is mass stranding's. Fortunately, this is a much less common situation, when multiple dead animals turn up in a single location. Mass stranding's often occurs when an individual whale gets into trouble and lets out a distress call, prompting the rest of the pod to follow and become beached alongside. That said, there is no definitive reason for mass stranding, and can be difficult to understand the cause.

Death by natural causes
Throughout history whales that have been found beached, have been attributed to natural and environmental factors. Natural causes can include factors such as poor or extreme weather conditions, animals being pushed into shore by heavy wave action. The degradation of physical conditions due to old age or illness, or problems arising whilst trying to give birth in shallow water. Dolphins especially can suffer in shallow water, when trying to hunt too closely to shore and accidently beach themselves. Even navigational errors can prove fatal if the animals get lost, finding themselves on the wrong end of a quickly receding tide.

Situations can also occur when individuals fail to correctly pick up very gently-sloping coastlines, and find themselves in much shallower water than they expect to be in. In several mass stranding 'hot-spots' such as Ocean Beach in Tasmania and Geography Bay in Western Australia, this hypothesis has been suggested. In both these places the slope is very slight and research has been conducted showing that the gentle slope may affect the echolocation used by the whales, enough to make it inaudible. This would therefore cause the animals to become confused and disorientated, often resulting in them becoming beached.

Another theory, however somewhat controversial, is one researched by former U.S. Geological Survey geologist Jim Berkland which states that radical changes in the Earth's magnetic field, which occur just prior to earthquakes, are to blame for increases in beached whales. Berkland states that this interference affects the ability of sea mammals and migratory birds to navigate, explaining the increase in mass stranding. Another possible explanation may be 'follow me stranding' were larger cetaceans follow smaller dolphins or porpoises into shallower waters. Large whales often use smaller ones to locate prey, following the pods. If a combination of strong tidal flow and strange seabed topography is encountered, the larger animals can be at a much higher risk of finding themselves become trapped, and getting beached.

Orca, who predate on both Dolphins and Porpoises, will also follow them into shallower waters. However, Orca stranding are actually very rare. Even when compared with the frequency of stranding in dolphins. It seems that Orca have some sort of strategy for avoiding beaching. Orca, being more intelligent and powerful, have simply learned to operate more affectively in shallow water. It is also a possibility that Dolphins are chased in shore by Orca. The Dolphins either beach as a result of attempting to escape or are chased towards shore by orca, in an attempt to trap them. This alternative theory is perhaps strengthened when you consider the predatory tactics of the Orca of the Peninsula Valdes in Argentina. They are known to arrive at a specific time every year's, just as young Elephant seals are first taking to the water, not realizing the deadly threat's lurking. Orca thrust themselves completely out of the sea in pursuit, becoming very skilled at beaching themselves, and retreating back into sea on the next wave.

Man-made Causes
Man-made causes although not as common as Natural causes, are by far the more worrying reasons for whales becoming beached. The individual is often washed ashore dead, or near death. Reasons for this can be as a result of illness brought about by contamination and/or pollution. Oceanic pollution is distressingly high. It can affect whales by either directly harming them, or by harming their food sources. Diminished food supply causing starvation. Both of these factors can impede reproduction.

Other human factors can be caused by animals becoming entangled in fishing nets and ropes, either discarded or still being used. Whales often become trapped, leading them to drown, or entangled, inhibiting the ability of individual to swim or feed properly. Many countries have legislation to inhibit discarding of fishing nets/ropes, however other don't. With many species being extremely wide ranging and the potential for nets and ropes to travel great distances in oceanic currents, the stability of the legislation is limited.

A slightly less obvious way in which human activity is thought to lead to their beaching behaviours, is through naval activities. There is evidence that the very loud noises produced by anti-submarine sonar cause injuries to cetaceans, which may lead to beaching. Necropsies have also found internal injuries in large numbers of beached animals, and it has been suggested that the use of sonar may be responsible for these whale deaths. It has been proposed that the sonar can causes serious haemorrhaging by creating extremely loud and rapid pressure changes in the water. This has been backed up on several occasions when beaching of cetaceans has occurred briefly following Navy sonar exercises.

In March 2000, the US Navy accepted the responsibility for the deaths of 17 whales in the Bahamas, which seemed to have died as a result of acoustically-induced haemorrhaging around the ears. These injuries probably lead to the whales becoming disorientated and resulted in them becoming stranded.

Along similar lines, the most significant mass stranding of any type of cetaceans along the British coastline occurred in July 2008. The death of at least 26 dolphins were as a result of naval exercises using Mid-range sonar.

As well as internal haemorrhages, sonar can also lead to injury in whales by causing a form of decompression sickness. Exactly how the sonar is able to create the bubble formation which causes the bends, is not properly understood.

Conclusion
Although death by natural causes is more common and wide ranging, as far as the case of Lulu, it is more tragic as she died as a result of human interference. As previously mentioned, Orca standings are very unusual.

Her death was rare and caused by a factor which in an ideal world, could never have been caused.

To live without living?

"Don't you ever get the feeling that all your life is going by and you're not taking advantage of it?"

For the Human species life is stress filled and increasingly complex.

As for so many other species, life's objective is not simply involved in the most effective and efficient approach to carry out our daily functions and processes so that we might continue on successfully towards the next day without dying. There is of course much more to our life than warding off death.

Modern day life can be a pain! There are too many people, there are too many cars and too many buildings, too much noise and there aren't enough quiet places. It's hard to even go for a nice quiet walk without being interrupted by someone's dog/child, a plane flying overhead or your phone bleeping. One of the reasons I enjoy working as a naturalist is it is one is able to experience some of the worlds truly quiet places. The top of a hill or the middle of a forest with not a soul for miles around. The sense of isolation and freedom can be somewhat liberating to the soul. A lot of people may not feel this way, although I'm certain that many do.

The first few lines of the poem beneath are feature in the movie, Into the Wild, which is based on the true story of a young man named Christopher McCandless. He dies in the wilds of Alaska while seeking adventure, searching for himself, and probing for true meaning in his existence.

"There is a pleasure in the pathless woods,
There is a rapture on the lonely shore,
There is society, where none intrudes,
By the deep sea, and music in its roar:
I love not man the less, but Nature more,
From these our interviews, in which I steal
From all I may be, or have been before,
To mingle with the Universe, and feel
What I can ne'er express, yet cannot all conceal."

Lord Byron from Childe Harold, Canto iv, Verse 178

Byron describes his favourite pass time, walking alone in a untouched woodland. The pleasure he can gain from listening to bird calls and picking wild flowers and spotting the occasional deer or rabbit. He was no naturalist or scientist but nothing gave him more enjoyment (in his famously busy life) than nature in its primary form. Whether it be staring out towards a seemingly endless ocean horizon, up at the starry night sky or examining a bee's hive or ant's nest; nature contains colossal majesty and petite society.

"Come forth into the light of things, let nature be your teacher"

-William Wordsworth

The diversions of the natural world are endless, so much so that I feel they give a life who enjoys them more substance, making any man woman or child much the richer for their place in it. As the much celebrated American author Henry David Thoreau wrote:

"I went to the woods because I wished to live deliberately, to front only the essential facts of life, and see if I could not learn what it had to teach, and not, when I came to die, discover that I had not lived..."

 from Walden,or; a life in the woods

I am not meaning to imply that people that have no interest in the natural world cannot live a fulfilling life. I am however suggesting that these people are missing out. There is nothing in my opinion that we are so innately predisposed to be so deeply and emotionally drawn towards. Life has never been so complex and stressful so as we beat on, boats against the current borne back ceaselessly into the past, an attachment to something archaic and elemental has never been so valuable. As the person who can find contentment in nature will never go unhappy and can be so very quickly and easily gratified in its wonders.

"He is richest who is content with the least, for content is the wealth of Nature" - Socrates

Myth and legend

"A myth... is a metaphor for a mystery beyond human comprehension. It is a comparison that helps us understand, by analogy, some aspect of our mysterious selves."

Christopher Vogler

Besides wandering in the woods, one of the other ways I enjoy to while away time is walking along a beach or seashore looking for things which catch my interest. Whether it be flotsam and jetsam* or animals in rock-pools or on the beach, the high tides and stormy seas can bring in many interesting things. Once I found over 50 Razorbills (a type of seabird) washed up on the beach as a result of a storm at sea. A tragedy but only an incredible sight.

One of my favourite things I have found however was pictured below, it's a plastic bottle covered in Goose Barnacles.

Goose Barnacles are amazing creature and they a have a funny story behind them which is the subject of this short chapter.

Back in the dark ages, before people knew much about nature and the natural world, they were unaware of bird migration. It was believed that the Barnacle Goose, developed from the Goose Barnacle. This was assumed as they were never seen nesting in places that were familiar at the time (they breed/nest in the North Atlantic.)

The confusion was prompted by the similarities in colour and shape. It was quoted in an old text...

"Nature produces against Nature in the most extraordinary way. They are like marsh geese but somewhat smaller. They are produced from fir timber tossed along the sea, and are at first like gum. Afterwards they hang down by their beaks as if they were a seaweed attached to the timber, and are surrounded by shells in order to grow more freely. Having thus in process of time been clothed with a strong coat of feathers, they either fall into the water or fly freely away into the air. They derived their food and growth from the sap of the wood or from the sea, by a secret and most wonderful process of alimentation. I have frequently seen, with my own eyes, more than a thousand of these small bodies of birds, hanging down on the sea-shore from one piece of timber, enclosed in their shells, and already formed. They do not breed and lay eggs like other birds, nor do they ever hatch any eggs, nor do they seem to build nests in any corner of the earth."

The barnacles were assumed to be like eggs which became attached by unknown means to wood before they fell in the water, turned into birds and few away. The Welsh monk, Giraldus Cambrensis, made this claim in his 'Topographia Hiberniae'. Since barnacle geese were thought to be "neither flesh, nor born of flesh", they were allowed to be eaten on days when eating meat was forbidden by religion.

Most interesting for me and the reason I found this discovery so exciting is the fact that goose barnacle do not live in the UK. They are a species of tropical and sub-tropical regions. The only reason that anyone in the UK became so familiar with goose barnacle was due to our seafaring history. We would sale to Cuba (for example) and these would become attached to the hull. Also of course they would commonly attach to pieces of flotsam/jetsam, as in this case.

In my head I was imagining somebody in Mexico or Florida, relaxing on a beach and enjoying a refreshing drink. The bottle is left on the beach and ends up being caught by the next high tide. Maybe there is a violent winds or a storm or cyclone that night and the bottle is set adrift, driven by the wind for hundreds of miles it is eventually caught up in the gulf stream and on its way towards the UK. It would take many years for a bottle to make it across the Atlantic but think of all it could see in that time. It would become a floating reef, tiny fish would swarm it for any sort of shelter they can find, which would attract larger fish, which would attract even greater pelagic predators. During its journey thousands of sharks will have swam under it. Not only sharks but it will have seen whales, sea turtles, marlin, everything! It will have endured countless raging storms and sat under infinite starry sky's, before becoming wedged in a rocky crevice on the west coast of Scotland.

*flotsam and jetsam are maritime terms, flotsam being wreckage that fell of a boat, jetsam being cargo that was deliberately thrown from a boat (ie. jettison) and washed ashore.

The Science of Non-Science

"What a man believes upon grossly insufficient evidence is an index into his desires, desires of which he himself is often unconscious"

Bertrand Russell

Living in the 21st century is great. We, as a species, have never had life so easy. Live is good for the majority of us, when compared to the life lived by our forefathers. Thanks to science and technology, we live comfortable lives with very few moments of unpleasantness, at least in the developed world. Science has given us vastly better healthcare, medication, sanitation, transportation, personal beautification... you get the point. So why now are so many people turning their backs on science, turning to non-scientific means for help in their stressful lives?

What I am referring to is pseudo-scientific products, and it is bane of many a scientist and medical professional's life. The problem is that pseudo-science claims to be science, and to most people is indistinguishable from science.

The major differences between pseudoscientific practices and products, and ones that have been validated by science, is that the former don't work. Pseudo-science is often designed to take advantage of the ignorance of the average lay-person towards science, and is highly effective in doing so. By using scientific language and throwing in the occasional fashionable buzz-word such as "organic" or "rejuvenating" They have found themselves a huge niche, and are especially popular with people who have been labelled the "more money than sense" market. Unfortunately for these people they also don't work, so whether it's a pill or a whole course of treatments, if it is a pseudo-science it also going to have only a pseudo effect.

How to spot pseudo-science?
I must say that I have not been entirely fair so far in my explanations and estimations of pseudo-science. I have been somewhat derogatory, which is wrong. As I should explain, any pseudoscience if to ever be experimentally tested and methodological standards were upheld, it would leap from the realms of mere pseudo-science and become science. And this does happen!

The basic standard for something to qualify as pseudo-science, is if it is presented to people as a being a consistent with facts, but actually fails to backup these facts. By either ignoring scientific research, or by the misuse of scientific method.

It can sometimes be tricky to tell at a first glance, or even after careful consideration whether something is true science or pseudo-science. By its very design, it is often intentionally misleading. There are a number of ways that are very useful in identifying pseudo-science from true science however, which I will discuss below.

Exaggerated, untestable or vague claims

When reviewing any research that might have been conducted, you might notice very basic and simple things missing from the tests such as the omission of effective controls, such as placebo and double-blind cross over test. Both of these are standard test in ever the simplest of experiment and their omission is a major red flag. Any research conducted without this sort of control, may have well not have been conducted at all, as its results are useless. This leads on to one of the other major red flags of pseudo-science, very often it not made clear how an independent researcher might go about carrying out the research on their own. Variables are kept secret and parts are hidden which would make it impossible to reproduce to research, which of course is an essential part of the scientific method.

Other major red flags include the use of vague or dumbed down language, rather than the precise and accurate scientific language and measurements that you would normally expect. On the other hand, the polar opposite is also used to try and influence the effect on the lay-person. The use of overly complex and technical/scientific language is also used (often incorrectly) to try and give the pseudo-science at least a science-y sort of look. This is often a very effective sales technique, as it can be put on the packaging or advertising of a product. It doesn't even have to be read or understood by a person, the mere presence of those big daunting looking words is enough to make people think, "This sounds like it might work". For example, the use of 'aqua' instead of water is common, I have even heard of 'dihydrogen monoxide' being used as a substitute for water. It's not incorrect, but it is ridiculously over the top, and put there with the single minded purpose of fulling people

Evidence, selection and proof

Another red flag to look out for when thinking about pseudo-science are testimonial and anecdotal evidence. This is testimonials, that might be printed on the back of the product saying how wonderful it is, or anecdotal evidence from people who have used the product before. I'm sure you are familiar with this sort of evidence, for example alleged reports of Bigfoot or UFO's. And you can take the testimonials on the back of products about as seriously as the police or scientific community take UFO and Bigfoot sightings. Testimonials and the like mean nothing, as true scientifically legit product does not advertise like this.

Scientific products advertise themselves through the results of its research, and effective products receive continued use through doctors monitoring the products success. If a product doesn't work for a particular person, which sometimes it doesn't, a doctor will assist on a new course of treatment until a suitable one is found. You certainly don't get that with pseudoscience, and usually end up with a medicine cabinet bursting with unused rubbish.

Another way to spot pseudo-science is though the selection bias often shown in research. It is a major part of the scientific method to present both data that supports your claims, and any that might not. Pseudo-sciences never show any off the conflicting research. They also pick and choose which data to include in research and as a result, seeming to produce perfect research. This is known as the 'selection effect' and it doesn't happen in true science, as the real world doesn't work like that, it has anomalies, random errors and other things that can mess up your data slightly.

In science, and to be honest most things in life, the burden of proof is on the person making the claim. If you claim that a simple glass of water can cure all the illnesses in the world, you have to be able to prove it! However, in the pseudo-scientific world the demand is reversed and very often pressure is put on the scientific community to prove a claim to be false or to not work. This tactic is used because it is universally impossible to prove a negative, and the pseudo-scientists are well aware of this fact. They use it to appear to put the scientist under a great deal of pressure by switching the burden of proof to the reverse side. For example, there is no evidence that homeopathy works after numerous trials. Yet, an advocate of homeopathy might say that science cannot prove that homeopathy doesn't work! This is true, as you cannot prove a negative, but is also certainly doesn't mean that it does work.

Many more...
I could go on and on about different ways of spotting pseudo-science, as they are really quite efficient at giving themselves away. Possibly it is because they are so often not scientists, and are not working within the scientific community, so often just don't know the correct way to behave.

An example of this often occurs when a pseudo-scientist receives criticism; they very often do not act like the majority of true scientists. A scientist would be upset of course, but they would be spurred into action and would quickly and thoroughly review their data, except their error and redo the experiment. Tight social groups often form within advocates of pseudo-sciences and they therefore take any criticism of their particular *beliefs* very personally, and will defend them with much vigour. Critics are seen as enemies and criticisms are seen, not as problems that need to be resolved, but as enemy bullets that need to be dodged.

Conclusion
Pseudo-sciences are rife, they are everywhere, from homeopathy in the NHS to Astrology in the daily newspaper, and they aren't going anywhere quickly. People believe in it two, I was at the pub the other night and a friend had a bruise, another friend said "oh, u need to take some arnica, it'll clear that right up" I interjected and said "well after a few days anyway" and the girls said "well yeah, it takes a few days" and I then said "and bruises usually vanish after a few days anyway don't they?" She looked at me and said, "Yeah they do, but I think it still works, you know?"

Luckily for everyone, most products do no active harm and the pseudo-scientific researchers continue to make no progress and be ignored by the scientific community. It's a terrible irony, if the pseudo-scientists ever want their work to be taken seriously and maybe one day become science, they need to accept science. The scientific method is designed to promote true science, and if something is of true merit it will be promoted from pseudo-science to science. However, it can only accomplish this one way ... and so far no pseudo-scientific product has taken that step.

Political logic

"Politicians were mostly people who'd had too little morals and ethics to stay lawyers."

George R R Martin

While I was writing this chapter, my home country was taking part an enormous event, the people of Scotland took part in what may turn out to be the most important democratic decision of our lives. The decision of whether to leave the UK and become and independent nation, free from the United Kingdom.

Personally I don't usually get involved in politics, I find all politicians hard to trust, it is an occupation of deceit, deception and unprincipled manoeuvrings. I don't recall the exact figures, but the vast majority of all politicians came from careers as lawyers or directors of big business, which possibly goes some way to explain why governments don't operate as fluidly and ethically as they should.

Dale Carnegie, the American writer of the bestselling book 'How to Win Friends and Influence People' once wrote; "Remember when you deal with people you are not dealing with creatures of logic but creatures of emotion."

Politicians are well aware of this fact, and well aware that it is easy to influence people by influencing their behaviour towards them. It doesn't matter if they have a weak side to argue or little evidence to support their side, if they can influence your feelings towards them then they are on to a winner.

As the date of The Independence Referendum approached we found the politicians on both sides becoming more and more aggressive to both attract people towards, and scare people away from Independence.

A lot of debating and arguments goes on in the run up to any vote, and in any debate there is almost always one side who's argument is weaker, or has a less robust and tenacious individual's arguing their position. The weak side is easy to spot (other than the sweating and shaking) as they will often use what are known as logical fallacies to argue their side in the place of actual facts. A logical fallacy is an argument that uses incorrect or unsound logic, therefore technically deeming their argument invalid. Some fallacies are committed intentionally (to manipulate or persuade by deception) others unintentionally due to carelessness or ignorance. Politicians use logical fallacies for a number of reasons, such as wanting to make an opponent look foolish, because they have no better answer to give or simply due to not knowing any better.

I often watch debates on YouTube because I enjoy to watch intellectuals such as the late Christopher Hitchens debating people and destroying people's arguments with intellect, knowledge and logic, it can be very entertaining. The first thing you often see with most of these debates is one of the debater's will often skip out onto the stage, all smiles and giggles and full of presupposed confidence. This is an intentional fallacy of logic, what is known as the Moral High Ground fallacy. Also known as the 'Holier Than Thou Fallacy' and is basically and attempt to seem to look like the good guy, it doesn't matter that only rubbish comes out of my mouth.

Every politician does it, they do it in the form of snobbery to try and win arguments and as opinions or general stances (for example, taking a stance for gay marriage or against war in the middle east) to try and win favour with the electorate. It's all a bluff, they might very well support gay marriage but it doesn't mean they can run a country. The intention is to make your opponent seem like the bad guy, and therefore lose favour with the electorate. But of course it is illogical to assume that just because you don't like a person that he is wrong.

A popular fallacious argument often uses by politicians the *Argumentum ad populem* fallacy also known as the 'Appeal to the people fallacy.' This applies that something is true simply because the majority of folk believe it to be true. Politicians will attempt to use such a fallacy often, such as the use of opinion polls, suggesting majority opinions are always the correct ones. This is also popular with religious types who like to claim that their religion is the true won simply because they have the largest number of believers.

Whilst on the topic of religious types, a common logical fallacy that they often use is involves shifting the Burden of Proof. As we all know, '*Onus probandi incumbit ei qui dicit, non ei qui negat*' or in English 'the necessity of proof always lies with the person who lays charges' or 'whoever says it sprays it'. What religious types often try and imply is that they are quite happy believing in their friend in the clouds without any proof of its existence and it is up to science to proof he doesn't exist. Until science can do that, god automatically exists. Not often to the same crazy levels, but politicians will do the same thing with the intent of putting their opponent under pressure or attempting to make them look foolish. The reality is that, to quote Carl Sagan "extraordinary claims require extraordinary evidence" and if you make a claim it is your responsibly to prove it true. This is one of the very bases of all science.

Back on the subject of politics, another popular logical fallacy is what is known as *'Post hoc ego propter hoc'* or 'after this, therefore because of this' which implies that if event B follows event A, then event B must have been caused by event A. You can see it make sense if B was a tsunami and A, an earthquake as there is a clear and obvious link between the two which is backed up by science. However, if B was record year for tourism in Scotland and A record breaking good summer weather, there is unlikely to be much evidence of a link. Politicians may use the fallacy in examples such as, since they have come into power the economy has revived and house prices risen. The link may just be a coincidence, or result of different or unrelated factors.

The 'Argument from Ignorance' fallacy (*argumentum ad ignorantiam*) is popular with politicians, it is the mistake of assuming something to be true just because it hasn't yet been proved not to be true. This fallacy was used extensively in the referendum with Scotland's currency, oil revenues, passport control, place in the EU... everything basically. If something hasn't been proven not to be the case, they will plain make something up and claim it to be fact, until it is shown to be otherwise with actual evidence. For example, claiming that Scotland's place in the EU would be affected, they don't have enough revenue to run the country or that if a new currency was creating it would be a disaster. There is no evidence that any of these claims are true but as there is an unknown factor, so the former is assumed to be true. It is similar to the Burden of Proof fallacy by implying ignorance of something as a position of power, however in the Argument of Ignorance, lack of knowledge is on both sides of the argument and a fear of the unknown.

Another common logical fallacy used in politics uses the 'Argument from Personal Incredulity' and this is another sneaky tactic which can often go unnoticed, to imply something to be false based on the fact that they can't imagine how it could be true. Basically the idea that of one can't see how something is going to work, it is a better tactic to staying with what we know. Obviously, just because a person can't envision a scenario in which something works, doesn't mean it doesn't work. It is often seen in interviews with the public might be asked their opinion of a specific topic, "Oh, I don't know much about that but surely it can't be a good" or "I'm not sure, but what's wrong with what we are doing just now."

One of the most common tactic used my politician is the Straw Man. It is carried out using a deliberate diversion away from the topic under debate, replacing it with a different topic that distracts from the argument under discussion. The aim is to distract from the real topic under debate by talking about something they have more knowledge about or that there is more interested in. This debating tactic is a standard for politicians when backed up against a wall, and has been famously and very effectively used throughout history.

The rest of the fallacies I shall discuss are all Red herring fallacies, which means intend to be misleading and make false arguments due to the lack of any real argument. The first is very popular in politics and the press, it is a type of *ad hominem* fallacy known as 'Poisoning the Well'. This is when you attack the arguer instead of the argument with the intention of discrediting them. The hope is if you can discredit them then nobody will take their point of view seriously. It's a pretty pathetic ploy but its common because it works, unfortunately.

The 'Appeal to Emotion' or *argumentum ad passions* is an attempt to manipulate the emotions of people towards one side, rather than win them over on evidence or facts. The most famous example of this being the 1952, "Checkers Speech" by Former US President Richard Nixon, who was under trail for spending £18,000 of campaign money. Instead of addressing the accusations he was confronting, talked about a dog as he has been given as a gift by a supporter. This is both an Appeal to Emotion and a Straw Man fallacy. Another example of the Appeal to Emotion is when an attempt to induce fear towards the other side and their views with the 'Appeal to Fear' fallacy, *argumentum in terrorem*. For example, suggesting the other side is pro-fracking or in support of the UK harbouring weapons of mass destruction, leaving us open to a nuclear attack. The reality is both sides may have their pros and cons but if you can induce fear, your argument is likely to become more favoured.

Politicians also like to exploit negative feelings that we may already have brewing inside, attempting to win favour using the 'Appeal to Spite' fallacy, *argumentum ad odium*. For example, implying that MPs are all rich and live in mansion's, spend the money they receive from the taxpayer on luxuries like expensive car's or holidays abroad. Whether it's true or not, it still has nothing to do with their competence as a politician or whether their side of an argument is correct, therefore it is a logical fallacy.

'Wishful Thinking', is an *argumentum ad passions* in which an ideal scenario or set of circumstances is laid out based on what would be ideal or pleasing instead what has been calculated based on evidence and reality. One of the major problems many people have with the campaign for Scottish Independence is that it may all be a wishful thinking fallacy. People would like the side one side is correct, so it is the correct decision to make.

A seemingly crazy put actually surprisingly popular fallacy is *Reductio ad Hitlerum* fallacy or the 'Reduction to Hitler' fallacy. It is an argument based upon comparing somebody to Hitler, or the Nazi party. Christians like to say that god exists because Hitler was an atheist, which he wasn't by the way. People such as Putin are often compared to Hitler in order to demonise him, therefore people automatically disagree with everything he says. Hitler can be interchangeable with any other person that is universally reviled, such a Margaret Thatcher. Politicians and their policies are compared to those of Thatcher in order to sway opinion. It's just about as low as an argument can get, maybe one above turning around and sticking out your tongue and shouting, "You smell!" You know any politician using this tactic is getting desperate.

The Science of a Sun Tan

"If I had to choose a religion, the sun as the universal giver of life would be my god"

Napoleon Bonaparte

During a very rare period of premature summer-like weather in Scotland, the topic of sun tans came into my head for possible discussion in this book. I figured that people might enjoy a better understanding of exactly what they are doing to themselves when basking in the heat. I'm not going to jump on the band-wagon and explain just how much damage you are doing yourself. In fact, I intend to do the opposite, as we never seem to hear about the benefits of tanning, which there are many.

Firstly, it's important to know what a tan actually is, as I'm sure many of people have never given it much thought. A sun tan is obviously the darkening of the skin, as a result of your skins natural response to extended exposure to the ultraviolet (UV) radiation.

What causes a sun tan?
The skin darkening is the result of an increase in the levels of the pigment Melanin, being released into your skin cells. This pigment is produced in separate cells called melanocytes, of which primary role is produce melanin. The role of melanin is preventing the excess absorption of UV radiation, which as we all know can be harmful. How much melanin is released affects how dark a tan you receive. How darks and easily you tan varies from person to person, and is dependent on genetics. As we all know, different people tan more easily or darker and some people don't tan at all.

When tanning, you are being exposed to two types of UV radiation UVA and UVB. UVA are the weaker (313-400nm wavelength) and UVB the stronger and more dangerous (280-315nm wavelength). The energy is indirectly proportionate to the wavelength. UVB and UVA also differ as UVB creates new melanin, whereas UVA creates the tan by combining melanin already present in the skin, with oxygen. UVA is the type that causes melanoma (the malignant growth of melanocytes and generation of several types of skin tumor) however actually causes skin-cancer less commonly than UVB. UVB is also thought to be responsible for moles, and is more likely to cause sunburn. However, it also causes the production of beneficial Vitamin D in the skin.

Health benefit of a sun tan

The skin produces vitamin D in response to sun exposure (in particular, UVB waves in the 285nm to 287nm range) which can be a health benefit for those with vitamin D deficiency. In 2002, an article was published claiming that 23,800 premature deaths occur in the US annually from cancer due to insufficient UVB exposures (apparently via vitamin D deficiency). This is higher than 8,800 deaths occurred from melanoma or squamous cell carcinoma, so the overall effect of sun tanning might be beneficial as well as significantly less dangerous than not enough sun exposure. Another research has estimated that 50,000–63,000 individuals in the United States and 19,000 - 25,000 in the UK die prematurely from cancer annually due to insufficient vitamin D. However, 10 to 15 minutes of sun exposure two times per week will provide adequate vitamin D. Further, sun exposure and tanning will not produce vitamin D when the sun is too low in the sky.

Another effect of vitamin D deficiency is osteomalacia, which can result in bone pain, difficulty in weight bearing and sometimes fractures. According to some 2007 research, sun exposure during childhood prevents multiple sclerosis later in life.

Ultraviolet radiation has other medical applications, in the treatment of skin conditions such as psoriasis and vitiligo. Sunshine is informally used as a short term way to treat or hide acne, but research shows that in the long term, acne worsens with sunlight exposure and safer treatments now exist.

For those who choose to tan, some dermatologists recommend the following preventative measures:

Make sure the sunscreen blocks both UVA and UVB rays. These types of sunscreens, called 'broad-spectrum sunscreens' contain more active ingredients. Ideally a sunscreen should also be hypoallergenic and non-comedogenic, so it doesn't cause a rash or clog the pores, which can cause acne.

Sunscreen needs to be applied thickly enough to make a difference. People often do not put on enough sunscreen to get the full SPF protection. Reapply sunscreen every 2 to 3 hours and after swimming or if sweating enough to make the lotion run. In direct sun, wear a sunscreen with a higher SPF (such as SPF 30). For playing sports the sunscreen should also be waterproof and sweat proof.

The rays of the sun are strongest between 10 am and 4 pm, so frequent shade breaks are recommended during these hours. Sun rays are stronger at higher elevations (mountains) and lower latitudes (near the equator).

One way to check how much sunlight you are receiving at any time is to check shadow length. If a person's shadow is shorter than their actual height, the risk of sunburn is much higher.

Wear a hat with a brim and anti-UV sunglasses which can provide almost 100% protection against ultraviolet radiation entering the eyes. Be aware that reflective surfaces like snow and water can greatly increase the amount of UV radiation to which the skin is exposed. Overall, all of these preventative measures are pretty self-explanatory, and any person with common sense would do most of them without thinking. Overall, sun is good for you in suitable levels and not dangerous if preventions are taken. It is certainly much healthier and far less dangerous than avoiding the sun altogether.

Is teleportation possible?

"Beam me up, Mr Scott!"

Out of interest, one of the world's most famous sentence fragments - "Beam me up, Scotty", were never actually said in any Star Trek episode or film. The closest Captain Kirk ever got was "Beam us up, Scotty."

Teleportation machines were the best piece of technology that they had on the USS Enterprise by a long shot. Imagine what the world would be like if teleportation was possible! Ok, so it would be the end of the motor, rail and air-transport industry but that would be a small price to pay (and in the long run probably a blessing) to have instantaneous travel to anywhere on earth. No more cars, buses, trains, planes and all the trouble and strife that goes along with them. The world would be a completely different place to live.

So is it possible?
I am definitely getting ahead of myself when describing a world with teleportation machines for transport, so first off is teleportation even possible?

Well firstly, there is more than one different type of proposed teleportation technique and some are more plausible than others. The simplest form of teleportation would be the simple transferral of data. Data derived from an object or organism would be precisely deconstructed at one end, sent to its destination instantaneously and reconstructed into its original form. Although one of the easiest to understand proposals for means of teleportation, it does have a number of highly technical issues which may be extremely difficult to resolve. The ability to deconstruct and reconstruct the object (such as a person) with enough accuracy that every single atom is exactly the same position at the far end as it was at the start, is an extremely complex and unlikely proposition.

A breakthrough in this field recently occurred where scientist where able to 'teleport' a single particle from one end of the laboratory to the other. Although this might sound like a step in the right direction, if we were to breakdown the human body into all of its individual particles and teleport each one separately it would take around 400 trillion years to complete. Who even knows how long that could take or even if the universe would still exist once you finally came all the way through the machine, personally I would rather walk.

There is also the argument that the person on the other side of the teleportation machine is not the same person as the one who hopped into it at the start, and in many ways this is correct. As the person at the 'receiving end' is actually made from merely the transferred data (like a copied file) of the person at the start, which is actually made of new matter already at the destination, therefore technically it is a new person. Whether or not more complex pieces of data such as memories and knowledge would be able to be transferred accurately is not at all certain. Quantum mechanics, including the Heisenberg uncertainty principle state that is impossible to make an exact copy of an object. The same principles are true for clones, which to be brutally honest is in many ways what the person at the other end would be.

Inter-dimensional travel

The only other type of teleportation that I can think of would be by somehow manipulating the fabric of space-time to our advantage. To be honest this is really starting to push-it a little bit in terms of a plausible explanation. Inter-dimensional teleportation is just about as far-fetched as it sounds and is (rather ironically) a favourite of science fiction authors. Moving increasingly away from the 'sci' and towards the 'fi' end of the spectrum, it involves leaving one physical universe and then re-entering it in a new location. This is achieved either by somehow taking advantage of parallel universes or through the use of some sort of 'worm-hole' or similar idea. Unfortunately, although it would solve many of the problems that arise with other teleportation techniques, it still remains impossible according to current scientific understands of just about everything.

Conclusion

The only conclusion that I can really make, based on at least my current scientific knowledge is that teleportation is just slightly more plausible than time travel. Technically it is just as difficult and would probably have just as many potentially catastrophic outcomes waiting to be revealed. However unlike with time travel, which I believe if it were possible there would be evidence of (such as trainer-prints in fossils and the like) teleportation doesn't suffer this problem. So in our minds at least it can still remain a possibility for the future, but not in any of our lifetimes I shouldn't expect. Unless of course you can find the Philosophers Stone, it's just about as likely.

The appearance of extra-terrestrial lifeforms

"In my official status, I cannot comment on ET contact. However, personally, I can assure you, we are not alone! "

Charles J. Camarda NASA Astronaut

As I explained in the previous chapter, although I will admit to being somewhat of a science geek you might find it slightly strange to learn that I have never been a great fan of Science Fiction. I am not really very much into Star Wars, Star Trek, Dr Who or any of these types of shows. I know entirely sure why I've have never enjoyed Sci-Fi, and it is simply because I am a scientist. I am easily annoyed by scientific inaccuracy or things that I know not to be possible (it the same reason that I don't find horror movie scary). Maybe I simply have too much respect for science.

Even from a young age, I have found it difficult to get into something that I couldn't believe. Chewbacca seemed to me to look like an extremely unlikely alien and I couldn't get my head around the fact that the forest of Endor, where the Ewoks lived looked exactly as if it were Earth. Stars Wars wasn't the only programme where everything was extremely familiar looking, though it is one of the worst offenders.

The reason for this is totally understandable whether it was down to lack of imagination or just down to limitations on the film crew. For example, a human actor has to get inside all of the alien costumes in Star Wars so upright standing, quadrupedal aliens were the only solution. Star-trek on the other hand were just plain lazy when designing the appearance of the alien creatures. Most of the aliens were nothing more but humans paint odd colour and miscellaneous pieces of rubber stuck on their foreheads. Even as a young child, I knew any potential alien life forms would not look like that.

So what's with the ego?
I do find it somewhat interesting how most alien life-forms, obviously made up by humans are always extremely human like in appearance (and quite often speak very good English). Think of all your favourite aliens in popular culture. Creatures such as E.T, Predator, the aliens from Alien, The Ewoks, and The Greys from X-files, all of these are

extremely similar looking. They all have a basic human-like body form including being standing upright, two arms, two legs, a torso, a head with eyes set above the nose and nose above the mouth, two ears on the head, a mouth with teeth, hands and fingers, feet and toes. Basically they all have a pretty unapologetically anthropomorphic body plan. So What? You might think that an alien would look like that, why not? Well you may not realise this but the chances of human shaped aliens existing, is about as likely as aliens from another planet speaking English as a first language. It just isn't going to happen.

This may sound to you like a pretty out-landish topic to be attempting to write about. How am I supposed to know what an alien life-form will look like? You would be correct if you were to call me full-hardy by attempting to make such a statement, so I don't intend to. What I can say however is that it is relatively easy for anyone to work out what aliens will not look like.

I can easily presume that alien visitors will not resemble a human in any way, shape or form. It is easy for me to say this as the specific conditions needed to create humanoid life forms is specific to earth. It has only occurred here, can only occur here and would only occur here once. Even if we were to start at the very beginning of life on Earth, right from the start with the same organisms. It would be impossible for all of the same life forms to evolve again, even if under the exact same environmental conditions. For this to occur it would involve much more than just every single individual animal to behave in exactly the same way as it did before. This would need to occur so that it could result in the same animals breeding together under the same conditions and resulting in the same new adaptation, which would eventually lead to the same new species evolving. The exact same new adaptations would need to occur again under the exact same environmental conditions, which would result in the same creation of the same new species. What is more, this would need to happen not just once, but continuously for thousands and thousands of generations of reproduction for hundreds of millions of years. As I'm sure you can understand, this is an unlikely ask. Even if you are to presume that certain traits and attributes are more likely to evolve and it is likely that similar types of animals might occur. Many times the evolution (such as humans) of species were heavily influenced by factors other than animal behaviour. Often things such as weather conditions (such as ices ages), which are controlled by the position of the sun in the solar system. The more we think about it becomes clear just how unique and special we all are.

It therefore becomes impossible that any alien life-form which has evolved on another planet, would or could resemble a human. It has taken over four billion years of uniquely occurring adaptations to produce a human, the same it true for all of earths species.

It is also important to understand that evolution does not have a set direction. Humans are not at the top of an evolutionary tree, everything doesn't evolve towards human form or indeed in any direction at all. To a sparrow, he is the pinnacle of evolution at the top of the tree, to an earthworm it is the greatest. If evolution were to start over again on earth there is nothing that would push it in the same direction, and make the same animals evolve. Nothing is pushing alien evolution towards a humanoid form. Apes moved out of the trees to eventually create humans due to the expansions of savannah like habitats, which in turn was the result of a change in a major ocean current on the other side of the world which altered the climate.

Conclusion
So, maybe now you can understand why I don't like Sci-Fi. Even the programmes that try to be as realistic as they can, still always have humanoid aliens, and it annoys me. I think it comes down to lack of imagination but any alien is always based on some sort of known life-form from earth. One thing I am certain of is that any alien life-form we ever meet will be unlike anything we have even seen on this tiny little blue and green planet. I can't imagine what they would look like either.

The evolution of the G- spot

"I heard women
have these things
called G-Spots
and if you
find a woman's G-Spot
it's like hitting the lottery
she'll love you forever"

Part of "G-Spot" by Dave Ochs

The topic of sexual arousal is always a hot area of discussion. During this chapter I am not going to discuss the dynamics of human sexual courtship, or indeed go into much detail about sex, although I will a little bit so don't stop reading now guys. I am going to talk about the G-spot, which has been a popular topic of discussion since it was first discovered by the German gynaecologist Ernst Grafenberg in 1944. In glossy magazines and sexual manuals around the world, there have been countless articles written about the mystical G-spot and its supposed wondrous workings. The fact of the matter is that science still cannot confirm that the G-spot even exists.

Erogenous error?
The idea of area of the female anatomy with the sole purpose of allowing for sexual pleasure entered popular culture after the publication of *The G Spot and Other Recent Discoveries About Human Sexuality* in 1982. It was heavily criticized almost immediately and has been ever since.

Ernst Grafenberg identified the location of the G-Spot as in the superior wall of the vagina, just behind the pubic bone. He also said that when stimulated correctly would produce a distinct orgasm, separate from a clitoral one. Both of these observations he was to become convinced of, solely based on the recounts of the experiences of patients. He was to publish his "ground-breaking" paper without any research to back up his claims, in an obscure medical journal. Today, proponents of the G-Spot are still criticised for giving too much attention to anecdotal evidence, and to questionable investigative methods.

Studies have been carried out into the existence of the G-Spot and the results don't look good for Grafenberg and the proponents of the G-Spot. As I mentioned earlier, the G-Spot is said to be located on the upper part of the vaginal wall, just a few inches inside and is said to be an area which is extremely sensitive to stimulation during sexual activity. Yet numerous scientific examinations of the vaginal wall have shown that there is no single area with a greater density of nerve endings, which would be needed to create this sort of sensation. The only thing that comes close is the urethral sponge, which both contains many nerve endings as well as erectile tissue.

What is the benefit of a G-Spot?

Other than not being able to locate the physiological site of the G-Spot, there are other problems brought up when conferring an organ upon the female anatomy with the sole purpose of pleasure. You need to have a good explanation to back up the lack of evidence. Any woman I'm sure will be the first to agree with me that when it comes to sex, men are much more straightforward than woman. What I mean by this is that during sex the man has a number of straightforward sexual targets which he aims for instinctually. The man wants easy access, control over the pace, pressure position and duration of the sexual act, rapid withdrawal once he is finished, finally he rolls over and forgets about what just happened. For men sex is purely about the act of sex and little else. This is why the prostitution industry is said to be the oldest in existence. As we all know, women do not work this way.

Women don't have the same orgasmic preconceptions about sex (although it is their aim I'm sure) and tend to become much more emotionally involved. Usually for a woman to achieve pleasure from sex the list of demands is considerably more extensive and many of them (at least the man might think) may not even be related to the act of sex. The environment should be optimum, small things such as where the acts is carried out, how dark or light it is or even whether or not there is music playing can be extremely important to how pleasurable a sexual experience the women is going to have. These things affect her psychologically and directly, as she becomes more aroused by the romance and other similar stimuli, as her body responds in anticipation.

When the couple actually physically meet, the woman demands extensive periods of foreplay and does not always want (or be physically able) to engage in the act straight away. Once inside the man's instinct drives his hips into a frenzy, yet the woman doesn't want that at all, she would much rather she could take control of the pace, pressure and duration.

Men and women differ greatly when it comes to sexual stimulation too. While a man will always orgasm during sex and feels little stimulation during the sexual act. A woman feels considerable more stimulation during sex but doesn't always orgasm, in fact some women never do. This explains why men often want to get finished, but women are so often left unsatisfied.

So what does this mean for the G-Spot?
As I have already explained, there is no evidence for the G-Spot and that people don't actually need a G-Spot to have sex. If this weren't already enough I am going to turn to 'old faithful' to imagine a world in which the G-Spot were a reality.

Often in science when considering a hypothesis, you just have to stand up, drop everything, forget all that you have just read and look at what you have for exactly what it is. Very often when you do this it allows you a clear and unclouded view of whatever you are looking at. When we do exactly this, it allows a new approach to be taken and breakthroughs can often become clear.

In regards to the G-Spot the approach which has never really been properly considered is exactly why the G-spot might have evolved in the first place. As I have already said, it has been put forward as a female erogenous zone and somewhat of on/off button of pleasure during sex. Well, let us take a backwards step and look at it again, does it fulfill these functions? We have already said that it cannot be confirmed that the area even exists, but we will assume that it does for the time being.

The G-spot is said to be in the upper part of the vaginal wall, a few inches inside the vagina. It is not an area of the female anatomy that is easily accessible during normal sexual activity. It is most easily accessed using as position were the woman straddles the man facing backwards, the reverse cowgirl, as it is known. Although this position might bring her some pleasure it is awkward to carryout, uncomfortable for the man and worst of all will lead to sperm being misdirected from its target within the vagina. Clearly the G-spot (or at least our interpretation of it) and Natural Selection are at odds with one another. Why would the G-Spot be in such a silly and inaccessible place if it were meant to bring sexual pleasure to women? It just doesn't add up especially as women already have an organ for giving sexual pleasure that works very well, the clitoris. The clitoris (unlike the area said the harbour the G-Spot) is a tiny area crowded with thousands of highly sensitive nerve endings. It has the highest concentration of nerve endings anywhere on the human body, exactly what you would expect from something that had the soul function of providing sexual pleasure. There would be no need for nature to supply the female body with a second pleasure organ, especially an inefficient and inaccessible one.

So what is it for?
So if the G-Spot does exist, what is it for? Although the evidence for the existence of the G-Spot is still cloudy, some scientists have put forward hypothesis for alternative explanations for its existence. Leonard Shlain in his book, "Sex, time and power: how women's sexuality shaped human evolution" proposed that the G-Spot may be an aid to ease the pain of childbirth. As I mentioned in my chapter "The Science of Childbirth" it is uniquely torrid affair for humans. Shlain suggests that the G-Spot may aid to reduce the pain somewhat. He implies the G-Spot is perfectly positioned to be stimulated by the baby as it was being birthed. This motion would release endorphins into the mother's brain both calming her and giving her a euphoric feeling, helping her to complete the birth of her child with less perceived pain.

Conclusion
Whilst researching this topic it became apparent that the G-Spot was quiet a misunderstood place. It also seems that some women do have a G-Spot and some women don't? That in itself doesn't seem right. Surely either all women do or all women do, as all women have the same internal anatomy. As there is no actual concrete evidence for the G-Spot existing I think it can be concluded it doesn't. Perhaps some are just more sensitive down there than others. It seems the only logical conclusion to make.

Why do we get Chickenpox?

"A spot. A spot. Another spot.
Uh - oh! Chicken pox!
Under my shirt. Under my socks.
Itchy, itchy chicken pox"

Chickenpox is no fun. However, it is something that everyone has to deal with and is as much a part of childhood as Santa Claus, skinned knees and swimming lessons. I remember being very young when I caught it. I don't recall who I caught it from but I remember my twin brother had it at the same time as me, and I was unfortunately on holiday at the time. This meant that chicken pox ended up being double as upsetting for me, firstly as it ruined my family holiday to Majorca and secondly because it meant that it was completely impossible to escape from my annoying brother all week! Some of the earliest memories I have were of me and my brother sitting on the stone floor, staring out onto the bright sunny balcony whilst covered from head to toe in pink cream, and scratching! Interestingly enough, there is still very little (if any) scientific evidence to support the usage of calamine lotion. It doesn't really do anything, except to further embarrass the poor child.

So, what is Chickenpox?
Chickenpox or more specifically *Varicella zoster* virus (VZV) is an extremely contagious illness, which is caused by a primary infection with that virus. Chickenpox is more common in the winter and early spring but can be caught at any time of the year. Chickenpox is a classic childhood disease and is most common in individuals between 4 and 10 years old. It is not as common in children younger than school age as they are not in contact with as many children.

The way the disease usually appears begins with conjunctive and catarrhal symptoms (excessive mucus production) followed by stereotypical spots which will appear in 2 or 3 waves. Spots usually appear on the head and torso and will become extremely itchy, small open sores after a few days and will heal quickly, without scarring.

Chickenpox has quite an interesting history with the first description of the disease being accredited to Giovanni Filippo, who was a doctor in the 16th century. Shortly after this initial description in 1600, the English doctor Richard Morton also described what he thought at the time was a mild form of smallpox. it took until 1767 for the physician William Heberden to prove that chickenpox was different from smallpox.

The origins of the name chickenpox are not certain, but there are several possible explanations. It has been suggested that the term "chicken" was given to the disease as it was seen to be a less dangerous version of smallpox and therefore a "chicken" version of the pox. The term "pox" is a medieval word for curse. As some thought that chickenpox was a plague brought on to curse children through black magic. It has also been suggested that the name originates from the fact that the spots that appear sort of look as if the skin has been pecked at by chickens. Some have also suggested that it is named after chickpeas as the spots are similar in size to chickpea seeds.

How do you know you have Chickenpox?
Chickenpox will form a blister, and it is the fluid from these blisters which spreads the infection by means of direct contact or through the air, from the cough or sneeze of infected person. A person who has chickenpox is contagious from 1-5 days before the spots appear, until all of the blisters have appeared and formed scabs (which can take another 5-10 days. Following contact with an infected person, it will take from 10-20 days for chickenpox to develop in a newly infected person.

The blister which is formed during chickenpox is never much fun. It starts off as a tiny (2-4mm) red *papule* which is irregular or rose shaped in appearance. A dew-drop shaped, clear vesicle then develops on top of the red papule. This characteristic lesion is often referred to as a "dew-drop on a rose petal" and is unique to chickenpox. Once the fluid has been formed for about 8 to 12 hours, it gets cloudy and the vesicle breaks leaving a crust which is no longer considered as contagious. After a week or so the crust will fall off which sometime leaves behind a scar which is also characteristic and instantly recognisable as a chickenpox scar.

It takes roughly a week for a single lesion to go through its complete cycle and new chickenpox will continue to appear every day for several days. As a result of this process it usually takes a week or so before new lesions stop appearing and existing one's crust over. People are considered highly contagious until all of their existing lesions have crusted over and should not be sent out into the general public.

People often don't know that they are infected with chickenpox before they can spread the virus to other people as they are capable of spreading the disease before any rash developed at all. Zoster (or Shingles) is a reactivation of chickenpox which is also another source of the virus for susceptible children and adults. It is actually possible to pass on the virus up to 2 days before the rash has even appeared on the skin, and up until the sores have totally crusted over (which will usually be completed 4-5 days after the rash first appears).

Pathophysiology of Chickenpox

The most common way in which chickenpox is spread is through the inhalation of droplets, which become airborne after leaving the infected host and being breathed in by a new host. As every parent is well aware, chickenpox spreads through schools extremely quickly, this is as a direct result of the airborne nature of the infection. Around 90% of people who have not had chickenpox will catch it on contact with an individual with the disease. The majority of people will catch the disease before they reach adulthood, however about 10% of young adults remain susceptible

After the initial inhalation of the infected droplets of fluid, the virus then infects the conjunctivae or mucosae of the host's upper respiratory tract. After 2-4 days the virus proliferates itself within the regional lymph nodes, which is then followed by primary viremia after between 4-6 days. What this basically means is that the virus settles down and replicates itself. After this secondary viremia, there is a viral invasion which is achieved by the virus spreading through the capillary endothelial cells and on the skin. The next stage is the production of the stereotypical vesicles which are formed by an infection of the cells of the malpighian layer which produces inter and intracellular edema (which causes the spots).

How does the immunity develop?

It is a well know factor of the development of both chickenpox and shingles that it can only develop once, and will not be caught more than once. So much so that it is quite common for parents to expose their children to the virus when then are young. The logic is that they will not catch it again when they are older when the infection can be much more serious. Infection in adults is more commonly linked to mortality due to the link with pneumonia, hepatitis and encephalitis (inflammation of the brain) and is particularly dangerous in pregnant

women. Pneumonia is in particular related to chickenpox with pregnant women, with up to 10% of individuals developing the disease.
Things such as bacterial infections of the spots are also relatively common and can often develop in to diseases such as impetigo (superficial bacterial skin infection) or cellulitis (infection of the deep subcutaneous tissue). Other, slightly rarer diseases sometimes linked to chickenpox include myocarditis (heart muscle inflammation) and glomerulonerphritis (a type of kidney disease).
For these reasons it is often said that it is best to get chickenpox when you are young, as you are not very likely to catch it again as immunity usually builds up. However, it does not always develop and it is actually possible to catch chickenpox more than once, especially if it has been a very long time since you last had it. Being exposes to the VZV is usually enough to initiate the production of the range of immunoglobin antibodies, which are necessary for the development of the immunity and it will usually persist for life. Other types of immune responses are also important for helping to limit the chance of catching chickenpox again, such as cell-mediated immune responses.

Treatment of chickenpox
One of the earliest of all the memories of my childhood includes the topical application of calamine lotion. Applied to chickenpox as a zinc oxide based topical barrier it is very commonly used, despite the fact that there is no evidence that it is any use at anything other than making you pink. However, it does have a very good safety record and is quite likely to help to maintain clean skin, which is very important to reduce to chances of secondary bacterial infection which is the source of all the complications which can be linked to chickenpox.
Other than that there really is not an awful lot which can be done to treat somebody with chickenpox. Certain individuals whom may be at particular risk if they were to catch chickenpox, can receive a chickenpox immunoglobin prior to the onset of the disease. Such individuals include people with weak immune systems, pregnant individuals or new born children as they are much more likely to suffer from secondary disease.

Other than that it is mainly just the prevention of infection which is of most importance and antiviral drugs can be beneficial if an infection develops in otherwise healthy adults. Also individuals who have severe eczema can be at greater risk of diseases caused by secondary infections and should be treated with anti-viral drugs as a general rule.

Conclusion

Once somebody has had chickenpox, nearly all of them will have developed the immunity. However not everyone does, and it is possible to get chickenpox more than once. That said, the virus usually reappears as shingles instead, which cannot be passed on to other people. Chickenpox is basically a disease which just has to be suffered through. There isn't really a cure and there isn't very much that can be done to treat the symptoms. Calamine lotion is still a popular treatment and remains the most common treatment for the spots. Possibly due to the fact that it is quite a successful barrier to infection and is really quite soothing when applied to the skin. I like to think it has something more to do with lovely pale pink colour, that's certainly why I liked it.

Why people who live at sea level age more slowly?

"Age is an issue of mind over matter. If you don't mind, it doesn't matter."
Mark Twain

I live at sea level; actually I live within a stone throw of the sea in a town on the West coast of Scotland called Greenock. It is important that I make it very clear that I am not for a second suggesting that people in Greenock are healthier or live longer or anything along those lines. The sad fact can be made reasonably axiomatic by the simple act of walking down the Greenock high street. My fellow "Greenockians" are not the healthiest of peoples. For example, recent government statistics showed the Greenock area has the fifth highest number of deaths per year due to alcohol related disease in the entire UK. So being at sea level certainly doesn't lengthen life in that sense.

That is not what I am going to discuss, besides I'm sure people must be bored to death hearing how unhealthy they are, and I don't intend to join in. What I do intend to discuss is much more interesting and I guarantee you will not have heard this one in any of your glossy, lifestyle magazine, which are so numerous nowadays as to take up an entire wall of shelves in my local super market (they have two science magazines).

Time is NOT constant
Theoretical physics is what I want to talk about. I hope I didn't just scare you off, because it is actually a quite simple and wonderfully interesting topic and I promise not to bore you.

People who live in Greenock and in truth all people who live at sea-level actually age more slowly than people who live higher up in the topography. This has nothing to do with what you are possibly thinking, but is actually to do with the passing of time itself. Time literally passes more slowly the closer you are to the earth, and of course by living at sea-level you cannot easily get much closer to the earth.

If you consider a pair of twins for a moment and imagine that one went to live on the top of a mountain, whilst the other stayed living at sea level. The twin on the top of the mountain would age faster than the one at sea level and if they met up again the twin who had been living at sea level would literally be younger than the other twin, as less time will have passed. In this case the difference would be far too small to make any noticeable difference. However, if one twin were to go on a long trip, at near the speed of light on a spaceship and then returned to Earth, they would have aged very little whilst the other twin would be significantly older, maybe even dead. This is because literally less time would have gone past whilst they were in space travelling at near the speed of light. That might have just confused you as surely the twin in space is much further away from earth. The key is that he was travelling at near the speed of light, and why is important will become clear is a moment.

A "Relative" explanation
How this is possible was first discovered by that most famous of all theoretical physicists, Albert Einstein. As part of his (and possibly anybody's) greatest piece of work, The Theory of General Relativity, he showed that time should appear to run slower near any massive body like the Earth. The reason this happens is because there is an important relationship between the energy of light and its frequency, which means that as the energy increases so too does the frequency. Therefore, as light travels up in the Gravitational field of the Earth it loses energy and therefore also its frequency decreases. This means that there is an increase in the length of time between each wave of light. This increases and continues to increase the further you move away from the Earth. Basically as you move away from the Earth, the greater the distance and therefore the length of time between each wave crest. This means in a way; time is taking a more or less time to occur dependant on how close you are to the Earth. To anyone who was high up it would appear that things happening lower down (closer to Earth) were taking longer to happen as space and time do not remain constant.

This may seem farfetched and quite hard to believe but it has actually been tested experimentally. In 1962 a pair of extremely accurate clocks were brought together and set to the exact same time, down to the accuracy of mere fractions of a fraction of a second. One was then mounted at the top, and the other at the bottom of a very high water tower. After a period of time it was discovered that the clock at the bottom (nearer to the Earth) was indeed running slower, in exact agreement with general relativity. The difference between these two clocks is actually of great practical importance in today's society of air travel and satellite technology as it becomes impossible to ignore these time differences. If one was to ignore the fact that time runs at different speeds, things such as satellite navigation would constantly be wrong by many miles, which could end up being extremely dangerous as well as more than a little inconvenient.

So what does this mean?
Until fairly recently (1915 to be exact) time was thought to be fixed and unaffected by anything that was happening in it. However general relativity brought an end to that, and showed us that both space and time not only affected but also were also affected by everything in the universe.

This new knowledge of time and space led on to scientists having a much greater understanding of the universe and how it works. Now that it was clear that space and time were dynamic quantities, it could be understood that this meant that the entire universe was actually a dynamic and ever moving place. It had to be expanding outward and ever increasing in size, which meant that it must have had a start and also must have an end. This idea was not popular with many scientists to begin with (as it wreaked of divine intervention) but as people started to better understand The Big Bang, they realised it actually made the idea of god look unnecessary. Of course it is possible that god made the Big Bang occur, but as time did not exist before the Big Bang and neither did the Universe, it makes it hard for god to have existed. I apologise if you don't fully understand, to be honest I don't properly understand it and I have deliberately excluded a lot of important detail in an attempt to not overly confuse anyone (including myself). I also apologise to any Theoretical Physicists, Cosmologists, Astronomers or any other people who are much cleverer than me, for any errors I may have made. I am but a humble Biologist, who doesn't even have a PhD! However very clever people do actually spend their time trying to solve these sorts of problems, and why not as they are some of the most fundamental and mind-blowing problems in the universe. And what have they managed to discover with all their tinkering with ultrasensitive atomic clocks? Well we can now calculate if you lived just 1 foot above someone else, over the course of 79 years you would end up being 90 billionths of a second older.

Why do we have an Appendix?

"The modern king has become a vermiform appendix - useless when quiet, when obtrusive in danger of removal"

Austin O'Malley

Nobody seems to like the appendix terribly much, doctors have no qualms in whipping it out of your body and even in my opinion the greatest person to ever live, Charles Darwin didn't think much of it. Darwin claimed that the appendix was nothing more than an ancestral, vestigial structure which had no absolute purpose in human beings. Even the first anatomists to describe the physiology of the appendix didn't seem to think very highly of the organ either. This becomes more obvious when you consider the proper and full name, which is the vermiform appendix. The word "vermiform" means in Latin "worm-like in appearance" which is due to its long and slightly segmented look.

The appendix is often the butt of jokes. Comedian Jimmy Carr once said, "Ah, the appendix, isn't that an extra bit of stuff added at the end of a body after it has been finished?" He was of course referring to an appendix as a supplemental informative section, added at the end of a book which gives extra information, updates or adds any corrections to the main work.

The Function of the appendix
So what the purpose of the vermiform appendix? The religious type might assume the appendix must have a purpose, otherwise why would god have put it inside us? However, it is common for an animal to retain parts of the body which that particular species no longer need, but a closely related species did need. Evolution allows the organs to be retained as it has not been put at advantage or disadvantage to have retained it.

There are many examples of this throughout the animal kingdom such as the wings of flightless birds like the Ostrich, the fact that whales have tiny back legs still attached to their skeletons or even the fact that male mammals have nipples.

The appendix is not a true example of this, as it is in fact a fully functional organ with a number of useful functions. Given its proneness to sometimes cause death via infection and the seemingly impeccable health of people who have their appendix removed, the biological function of the appendix has been somewhat of a mystery for some time. There have even been a number of people born without an appendix, with no reports of any sort of physiological impairment suffered.

It is not really much of a surprise then that we have always believed that the appendix has no purpose, and is indeed a vestigial organ. As I mentioned earlier, one of my heroes Charles Darwin thought the appendix was a vestigial organ, believing it to have ancestral purpose only. Darwin suggested the appendix had a use in the digestion of leaf matter when we were leaf-eating primates. Few mammals other than the Primates have an appendix and its level of development decreases from Human Primates with the least, to Old World Monkey Primates with the most developed appendix. This evidence along with a great deal of other research to support it, suggests that as per usual Darwin was correct about this. And that the appendix is a vestigial organ in humans which originally evolved in Monkey-like Primates.

Immune use of the appendix

Recent evidence however has suggested that the appendix also has a potential immune function, which seems to contradict the explanation of the appendix as a vestigial organ. However, to say this would be to misunderstand the definition of the vestigial status in evolutionary theory. The true definition of a vestigial organ or limb does not necessarily mean that the organ/limb has lost all function. The vestigial organ/limb can also develop new functions often completely separate from the original use. So Darwin was actually both correct and incorrect about the appendix.

The immune system use of the appendix was first suggested by Professor Loren G Martin of Oklahoma State University. Professor Martin suggests that the appendix is useful in both adult and the foetal stages of human development. The endocrine cells (hormonal cells) which have been found in the appendix of young foetuses have been shown to have an important role in biological control of the foetus. In adults the appendix has been suggested to have an important function in the immune system as it is rich in lymphoid cells, which are important in fighting infection. It has been suggested that it has a role in both the manufacturing of hormone during the development of human foetuses and also a function in 'training' the immune system. It achieves this immune purpose by exposing the body to antigens, allowing it the ability to produce antibodies (used by the immune system to fight bacteria/viruses etc). The appendix is also now thought to be of importance in the recovery from diarrhoea. The levels of fatality from diarrhoea are extremely low in developing countries when compared to the huge number of children who suffer from it each year. The high recovery rate is thought to be due to the fact that appendixes are not removed in these countries, resulting in improved immune systems. The importance of the appendix is now thought to be such, that it is no longer removed during surgical procedures as it used to be - just in case.

Most recent explanation of function

Linked to the appendixes useful function in the immune system, William Parker and Randy Bollinger, of Duke University, suggested that the appendix had a further function associated to the lymphoid tissue of the gut. This tissue is known to have an extremely important purpose in the growth of beneficial intestinal bacteria which are vital when illness strikes. The appendixes proximity to this important immune tissue led these researchers to suggest that the shape and architecture of the appendix allow it to act as a "safe haven" for these bacteria. As the useful bacteria is often flushed out of the body by illness, it was proposed that it acts as a site were the bacteria could congregate as well as avoid being washed away, allowing stocks of beneficial bacteria to be kept high. This is the similar type of bacteria which you are ingesting when consuming Yakult or Actimel or similar, and that is in fact all you do when you take these probiotics. You increase the amount of bacteria which are already there in significant levels; although it is still not certain how affectively these drink work. They have been shown to have some beneficial uses in the treatment of certain specific infections and diarrhoea, but they were consumed in extremely high levels under hospital conditions. Personally I would trust your appendix over any probiotic drink! The appendix has developed over millions of years of evolution, has many useful functions and is supported by credible and significant scientific research, unlike probiotic drinks despite blatant advertising claims suggesting otherwise.

What you might not know about evolution

"You think you're so clever but I'll live for ever. You're just a survival machine"

Richard Dawkins in The Selfish Gene

As a biologist my principle passion is the study of evolution and as a result one of my most common annoyances (of which I admit I have a few) is people misunderstanding and misinterpreting the processes of evolution. It is not really their fault to be fair, I didn't realise it until I started university but the way in which we are taught evolution at school leaves it open to many errors, even when you do understand its basic principles. The purpose of this brief chapter (and I had to keep it brief or it would have never ended) is to address just a few of the most important misunderstandings about evolutions. Selected are the ones that I have to explain to people the most frequently.

It's an interesting fact that Darwin did not once use the word evolution in his book The Origin of Species. It is also an interesting point that he did not talk much about the origins of many species either. When you give the book its full title however, which is The Origin of Species by Means of Natural Selection, or The Preservation of Favourable Races in the Struggle for Life, it becomes clearer what exactly Darwin was discussing. This was the introduction of the theory that populations evolve over the course of generations through a process of natural selection.

In its time it was seen by some as a controversial theory, it still is by the people who prefer the comforts of religious belief. Yet many saw it as welcome payoff to a belief system they had long held. Only until then, they couldn't justify their non-theistic beliefs with an alternative to super-natural creation.

One of the principle misunderstandings about the Theory of Evolution is that it is "just a theory." This mainly comes about due to the misuse of the word theory in common culture. In most of the sciences) except physics which rather pretentiously uses the term 'Law' instead) a theory is considered the closest anything can be seen as an absolute truth. What people can't consider a theory, is in sciences know as a hypothesis. To prove a theory takes a large number of repeated experiments under controlled conditions, all coming to the same conditions. Thomas Huxley (AKA "Darwin's Bulldog) was once quoted saying "The great tragedy of science - the slaying of a beautiful hypothesis by an ugly fact." Known for his aggressive defence of Darwin's theories at the famous 1860 debate, he was renowned for his sharp sarcastic wit. I am not certain what he is referring to in the above quotation, though it could as just as easily have been targeted at the religious leaders he was debating.

Evolution holds the same level of scientific standing as gravity or the fact that the earth revolves around the sun, in its factual accuracy. A fact that the fields of palaeontology, biological, biochemistry etc, independently confirm and show commonalities without reasonable refute that the structure of the genetic code of all living organisms, including humans, clearly show a common primordial origin.

Other simple misunderstandings
I am sure most people are aware of the basics of how evolution works, it is pretty much in the public consciousness by now and we are all taught it in school. We are however (in my experience at least) taught it very poorly. I doubt my biology teacher knows her Punctuated Equilibrium from her Peripatric Speciation, **and probably couldn't tell a Koinophilic species if it soiled on her dress.** As you can imagine, the process that has created all of the species that ever lived can get extremely complex in the detail, although incredibly simple in it basic principles. I am not going to delve into the complex depth of modern evolutionary theoretic however, I would like to correct a few of the common misconception about evolution.

Another misunderstanding about evolution is that it random, and relies on chance mutations. It is unfortunate that the definition for mutation in the biological world is again, different from that of the general public. Evolution does not work by random mutations, if it did it wouldn't work and we would still be bumping off each other in primordial soup. Evolution is the process of random mutations reinforced by non-random natural selection. Natural selection is the very opposite of random, it is in fact extremely selective and precise. When a mutation is favourable, natural selection allows it become more common, when it is not natural selection will not allow it to survive. With natural selection in operation, favourable new traits can evolve at an incredible pace. A fact that can be seen in the fact that bacteria and viruses continue to out-compete our ability to overcome them with drugs. Obviously, animals evolve much slower but the theory works the same.

Again, a common misunderstanding about evolution which is that evolution has been working towards creating humans, evolution doesn't have any direction. Evolution has not finished at humans and it will not end at humans. Many have the false idea that the whole of the evolutionary process leading to creating humans, as if the natural history of the planet was nothing but a prelude to the creation a bunch of lanky, feeble, hairless apes. Humans or *Homo sapiens sapiens* are just one species of all the living species which just happen to exist right now, and are all at exactly the same stage in the evolutionary history of the planet. As time goes on, all of the living branches will continue to evolve and break off into new species or will become extinct. We are not any different in that sense. We will die out and be replaced just as the species that came before us have, we aren't special and the world wasn't made for us.

On a similar point, humans did not evolve from chimpanzees. Chimps are at exactly the same evolutionary stage as we are. We do however share a recent common ancestor with chimps, a nearby branch on the evolutionary tree. This means that they are very similar to us both physically and genetically, mainly because there hasn't been enough time to evolve many major differences. We aren't related to chimps; we all have an ancestor who we share a 250,000th great-grand parent with a chimp, but that's as close as it gets.

The last example in this short list of common misunderstandings about evolution is that it can explain the origins of life. Evolution never has claimed to be able to say how life started or anything about what started life. Evolution can only explain what happened straight after life first appeared. How life began is still not 100% understood.

Currently, evolution is under attack from an old adversary, known as creation Science, or Intelligent Design. This is a phenomenon, mostly based America, that aims to baffle the layman with supposed scientific proof that "god did it." If you come across this, it is not science; it has no place in science.

What is a human body worth?

"If anything is sacred, the human body is sacred."

Walt Whitman

A few years ago, after being declined for a job application, was contemplating my worth. Whilst doing so I began to literally contemplate how much I might be worth. What I mean is, if a human body was broken down into all its individual chemical elements, how much would it be worth, in cold, hard cash?

To work this out it is first important to work out the chemical composition of the human body by weight. Ignoring some of the smaller, trace elements which aren't really important enough (monetarily) to include, there are really just ten main chemical elements making up the human body. Below is a list of these elements, with the percentage weight they make up.

Oxygen	65%
Carbon	18%
Hydrogen	10%
Nitrogen	3%
Calcium	1.5%
Phosphorus	1%
Potassium	0.35%
Sulphur	0.25%
Sodium	0.15%
Chlorine	0.15%

As you can see we are composed of a very high percentage of oxygen which is odd as life of Earth is often described as being carbon-based. This however is a reference to the carbon bond that is important in complex molecules, essential for life. This bond is vital for the bonds between carbon, hydrogen and nitrogen in particular, the next three on the list.

Now that we know what the human body is made of, and how much of each chemical can be recovered from a body, it is possible to work out how much a body is worth in terms of its elements. The next step is to work out how much, in weight, of each chemical can be taken from the average human body. As the average human weighs around 70kg, that is the total weight of chemical constituents that we have to work with. Baring this in mind, this means that by weight we have the following amounts of elements.

Oxygen	45.5kg
Carbon	12.6kg
Hydrogen	7kg
Nitrogen	2.1kg
Calcium	1.05kg
Phosphorus	0.7kg
Potassium	0.245kg
Sulphur	0.175kg
Sodium	0.105kg
Chlorine	0.105kg

Assuming that the chemicals extracted are of average quality, then the value of each chemical can be worked out using a chemical catalogue. Just as a chemistry teacher might order in new magnesium strips, all the chemicals that make up the human body can be bought wholesale.

I actually had quite a bit of difficulty getting my hands on a chemical catalogue. I contacted a number of universities and schools via email, only to be ignored and I tried to track them down on the internet to no avail. The closest I could get was antique chemical catalogues from the early 20th century, which seemed to be quite popular. So I borrowed one of these from the library, and decided I would work how much the human body was worth at least when the book was published.

When I added all of the chemicals together it makes for the total value of the chemical constituents of a human body to be about £650.

Doesn't sound like much, especially if you consider the astronomically high cost of organ transplants and similar surgeries to the lucky reciprocates of the life-saving organ. You might think that with the way the economy and money markets work that the value of a human body would be greater in the 21st century. However, you also must consider that the way that chemicals are mass produced today makes them much cheaper and less valuable.

The U.S. Bureau of Chemistry and Soils recently invested many a hard-earned American tax-payer's dollar in calculating the chemical and mineral composition of the human body. They calculated that when we total the monetary value of a human body it comes up with a total worth of only $4.50! That is actually including all the elements in our bodies and the value of the average person's skin if it was to be and sold as leather. Heaven forbid.

Shocking, eh? The truth is if you were to break down anything into its chemical constituents, whether it is a human, a car, a tree or a computer; it isn't worth much in monetary terms. However, when a product is valued for sale it's not just the cost of its part which make up the value. More important are what the product can do, how complex it is, how many processes and functions it can carry out, how much energy, effort and time it took to make and how much of a loss it would be if it were to be destroyed or lost. All of which would need to be considered when valuing a human life. If you were to break down a cow into its chemical elements it would be worth far more than a human. However, for the reasons I have just mentioned and many others, no sane human would value a cow's life over any human. Even though you literally can put a price on a human body, you can't put a price on a human life.

What is a hangover, and how can we cure it?

"good times came and went
broken-but not forgotten?
regurgitation"

The Hangover Haiku by Chris Gardner

Hangovers are not fun. Every single day, all over the planet, thousands of people wake up with a thumping sore head, a spinning bedroom and a toilet bowl that doesn't know what has hit it. But what exactly is a hangover, how is it caused and is there anything at all that can be done to reduce its effects?

What causes a hangover?
Well everyone knows that alcohol is the cause of a hangover but what is it about alcohol that makes us feel so awful the next day? Well alcohol is a diuretic, which means that it speeds up the rate at which water is lost from the body. This leads to an increased rate of dehydration, and it is the dehydration which leads to most of the unpleasant symptoms associated with a hangover.

The diuretic effects of alcohol cause the unquenchable thirst, pounding headache and that feeling of dizziness that we often get. The other rotten feelings we acquire such as sickness, indigestion and heartburn are caused by the way in which the booze affects the stomach lining. The tiredness is due to the fact that you have probably stayed up late into the night and will not have slept very deeply. This is because your body will use up a lot of energy trying to sort out the damage the alcohol has caused and you probably will not have eaten enough, or likely not very healthily if you did eat.

How to avoid a hangover
Not drinking any alcohol is the best way not to get a hangover...but let's be serious; if you want to drink you're not going to let the threat of a hangover put you off. Besides, pain has very poor memory, a survival strategy which has been majorly beneficial to our survival. Today it keeps mothers having multiple babies, just as it keeps people playing contact sports. You simply forget how much you didn't enjoy your last hangover.

Before you go out you drinking, you should always try your best to eat a decent sized, fat rich meal. By eating a meal high in fat, your stomach lining will be protected, as fat is digested slowly. Also by drinking a glass of milk it might help protect your stomach lining and slow alcohol absorption although for some people, drinking that much milk might just make you sick.

Try your best to alternate between alcoholic and non-alcoholic drinks (but not fizzy ones) as this limits the amount and speed at which alcohol enters your system. It also means you never have too much alcohol in your system at any one time and makes sure that you remain well hydrated.

If you don't want a hangover, then avoid fizzy drinks. Bubbles make you get drunk faster as the alcohol enters your bloodstream at an increased rate.

Lastly, whilst you are still out drinking try not to drink to excess, obviously if you drink stupid amounts of alcohol you're going to get a hangover regardless of how many precautions you take.

Once you have finished drinking and left the bar or nightclub you should consider walking home, if it is safe to do so. The fresh air will help reduce the effects of the hangover as well as help you to sober up faster. You should also endeavour to obtain a sports drink or some water, probably both actually. By drinking a pint of water it will help to re-hydrate your body before you go to bed. You could also try taking some vitamin C or at least drink a glass of orange juice, as even very small amounts of vitamin C will help to speed up the metabolism of the liver to get rid of the alcohol. Lastly, eat some cookies or toast or something similar, the lack of sugar is responsible for that wobbly feeling and they will also help settle your stomach.

Curing a Hangover
First things first, the hair of the dog tactic which is so whole-heartedly promoted by music festival goers, Ibiza faithful's and alcoholics alike this is not a good idea. Although you may feel better temporarily, you are only delaying the inevitable, setting yourself up for a much worst hangover, and causing yourself worse and possibly long-term damage.

Everyone has their own hangover cures. Personally I like to drink a bottle of sport drink and eat three or four packets of salt and vinegar crisps. I found this remedy out by accident but I worked out that the science behind it is that it restores your water, sugar salt balance. It works like a treat for me and my hangovers never last very long, sleeping all day long also helps.

The best idea is probably to just be well prepared. Have antacids for indigestion, paracetamol or ibroprofen for headaches and drink plenty of water. For breakfast have a few eggs, they contain high levels of cystenine which are known to help breakdown harmful chemicals that build up in the liver. A full English breakfast is a good idea, if you can stomach it, as the eggs, toast and the general fatty goodness will help you recover faster.

The effects of a hangover will usually be gone within 24hrs if you're unlucky. If you are lucky you will feel better by the evening and be ready to start all over again. Although... I didn't just say that.

Which came first, the chicken or the egg?

"The problem about the egg and the hen, which of them came first, was dragged into our talk, a difficult problem which gives investigators much trouble."

Mestrius Plutarchus, born 46AD

The above quotation is the first known commentary on the now famous question which has baffled many intellectuals of great weight since the Greek times. Just what did come first, the chicken or the egg?

The confusion comes from the fact that chickens obviously hatch from eggs, but in turn eggs are laid by chickens, which makes it hard to know which gave rise to which. At a glance it seems like a cycle of illogicalness which has caused confusion to some of the world's great brains since time immemorial. To the philosophers of old (who were the scientists of their time) it was a much deeper question than just the literal origins of wildfowl. It was a massively important question about life, the universe and everything.

Today with our greater understanding of the world and particularly as a biologist, the answer to the chicken/egg question is obvious. However, there are different ways in which you can approach the question which will evoke different outcomes.

Firstly, it is important to ask the question of whether the egg that we are talking about is a chicken egg or just an egg in the general sense. This might not sound like of much importance, but it is of vital as its answer will swing the answer one way or the other.

If we are talking specifically about a chicken's egg, then the answer is that the chicken came first. As with everything in life it, comes down to evolution. The modern domestic chicken is believed to have descended from another closely related species of jungle fowl. Domestication occurred after selective breeding in the same way as with dogs, cows, pigs etc were all bred.

So at some point (all thought it would be impossible to pin down or even notice that point) there was an egg from a domestic junglefowl, which contained the first chicken. There much have been a point where a chicken was inside the egg of a non-chicken. The first chicken would hardly be recognisably different from its parents but technically you would be able to say that it was a new species. However, is this actually a new chicken? Is it really a new species, to be honest it's a little bit of a grey area.

Genetically minor differences had occurred in the embryonic development of that chicken, after the formation of the egg. Even though the genetic events which set up the developmental changes may have started before the egg was fully formed, they did not really come into effect until later. Although it could be argued that they couldn't be stopped so may as well have been. I told you it was a grey area.

The most important point is that the parents were both not chickens, a non-chicken laid a non-chicken egg, and chicken hatched out of it. Therefore, the chicken came first.

So what if we are talking about just an egg in general. I think this is a considerably more likely, I don't think the ancient philosophers were obsessed with chickens. Chickens were just a common bird that were abundant at the place and time that the problems were first conceived, it could have just as easily been pigeons, gulls or crows. The chicken and egg was merely a convenient analogy used by philosophers for a much grander question. They were talking about the origins of life, but I am still talking about the origin of the egg. So, if we are not talking just about a chicken's egg, then which came first? How far back do eggs go?

Well the answer is simple and straightforward if we are not restricted to any particular species or timeline. The egg evolved long before birds, from invertebrates to fish, from reptiles to mammals, they all use eggs and have done since life was on Earth was quite a new phenomenon. I'm sure that for many people when we think of eggs, we always think of our feathers friends, the birds. Birds are actually one of the most recent of all the major animal groups to commonly lay eggs, which they acquired from their dinosaur ancestors. Reptiles, fish, invertebrates and even some mammals have been expert egg layers for millions of years before birds came on the scene and stole the spotlight.

Often branded as a great unknown question of science as if the answer is not and cannot ever be known, the chicken and egg paradox is really nothing more than a little bit of circular logic. Possibly it meant more to the ancient philosophers who had less scientific knowledge and much more time to waste debating about pointless topics.

It remains a dinner party favourite to this day, usually by someone who thinks he is going to silence the entire table by bringing up a philosophical topic. Well at least you can do your bit for science and humanity by putting them in their place

Why does the Hammerhead Shark, have a hammer shaped head?

"There's nothing in the sea this fish would fear. Other fish run from bigger things. That's their instinct. But this fish doesn't run from anything. He doesn't fear."

Peter Benchley

I have always had a fascination with sharks. The cold, dark and emotionless stare. The beautiful, silent and deadly character. But most of all, their reputation as a fierce, merciless killer and ruler of the ocean realm.

The hammerhead is a shark in which for many people, conjures up frightening images of a fearsome monster of the dark unknown depth of a world beyond their comprehension. Whether literal or symbolic, this perception is not accurate of course; of the nine species of the hammerhead (known as Sphyrnidae) they are all warm, shallow water species that generally pose very little threat to humans. In fact, scuba-diving with massive schools of several hundred hammerhead shark has become very popular in recent years.

All species from the smallest, the Bonnethead *Sphyrna tibruno*, which is only 3 feet long, to the largest the Great Hammerhead *Sphyrna mokarran,* which can grow up to a 20 feet, all have the unique hammer shaped skull, which gives the shark its common name.

There are a lot of ideas about why the hammerhead has this unique cephalofoil shaped head, ranging from the highly plausible to the just plain silly. One decisively fishy explanation tells of how the hammerhead will use its head like a hammer to club prey to death. This is not the chase as not only would it be an uneconomical way to feed, but as the head is made of cartilage, it would also be ineffective. Moreover, the eyes are positioned at the side of the head, so using the head as a hammer would involve smashing the prey with its eyes! A possible explanation for the misunderstanding of this behaviour may be in the way that the shark will shake its head violently and continuously when it has a hold of prey. This helps the teeth to tear through the flesh, so the shark to swallow a mouthful.

All of the other explanations of the hammerheads hammer head are indeed highly plausible and until recently the scientific community wasn't at all certain which was most likely.

Hammerhead Sharks are very large and bulky sharks, and as a result of this they are negatively buoyant (they sink). The shape of their head is thought to act like an underwater wing, to increase lift when swimming. Laboratory experiment on young hammerhead has confirmed that the head does act like a wing to help the shark stay up in the water, however this is likely not the reason it originally evolved.

As well as giving the hammerhead lift, the cephalofoil also improved the hydrodynamic efficiency of the shark. This cuts straight through the water creating less drag over the body of the shark. This allows the hammerhead to move extremely quickly, and create the shortest turning circle of any shark. Experimental data with young hammerheads has shown how the can make extremely fast and sharp turns, with their heads acting as a pivot point. This is an extremely important adaptation for hammerheads as it allows then to pursue and catch fast moving prey that other sharks may not be able to. Yet again however, this is likely only a secondary adaptation.

Sharks are famous amongst animals for their incredibly acute senses. The hammerhead shape allows the shark to have an even more accurate with is sense of smell than other sharks. As the nostrils are positioned wider apart, the shark was a wider and more accurate field of smell allowing it to locate smell more precisely. The widening of the nostrils came about as a side effect of the shape of the head.

The most impressive sense that sharks have, is a power which is totally alien to us. Sharks have the ability to sense electromagnetic currents and gradients. Sharks are able to harness this through specialized organs known as the Ampullae of Lorenzini. These organs contain hundreds of small jelly filled pits, all over the head and mouth and down the side of the body of the shark. This adaptation is primarily useful for catching prey. All animals create a tiny amount of electricity when it moves which the shark can detect. Due to their strange head shape, hammerheads have more Ampullae of Lorenzini than any other shark species, and as a result are significantly better at detecting electromagnetic currents. It has even been strongly suggested that they use this ability to navigate, using the earth's magnetic fields. This is a strong possibility to explain the head shape, however it is still not the best.

What is more important that seeing to a shallow water predator, better vision means that you stay alive as you are less likely to be eaten and you are more likely to find food. So it is not surprising when scientist finally concluded that the driving force behind the evolution of the shape of the hammerhead sharks head was to improve its vision.

A recent study has shown that hammerheads have excellent binocular vision. This is an important trait in all predators and is not really surprising to be honest. What was quite surprising was the fact that the odd positioning of the eyes allows the shark to see in almost 360 degrees. This means that the hammerhead can see in front, behind, above and below at all times, with fewer blind spots than any other species of shark. Dependant on the position of the eyes on the shark's head, it is possible to swim within sight of most species of shark without it seeing you, by staying in its blind spot. They general rule is if you can see its eyes it can see you, if you can't then it probably can't see you, though it likely still knows you are there. The hammerhead has a much wider range of vision, and it is believed it is this which encouraged the evolution of the hammerhead. The advantaged are easy to imagine, improved predator detection for young would allow more to survive to breeding age, and improved prey detection would mean that more adults would be able to remain fit a healthy. Both leading to improved breeding rates.

Conclusion
The hammerhead shark was once the most feared shark amongst mariners due to its odd, some thought grotesque appearance. Today however it is one of the most popular sharks to go diving with. Although they are still considered potentially dangerous and should not be underestimated, hammerheads are not responsible for a single recorded attack on humans. It could be said you are at more risk of being struck by lightning whilst scuba diving than attacked at being attacked by one of the sharks.

The reason I wrote this chapter is because hammerhead sharks can teach us a great lesson about science. There were so many hypotheses about the hammerhead's odd head. Originally it was assumed it couldn't see well, this was disproven. Then science discovered all its other wonderful properties and now the origins of the heads original divergence has been discovered. All through good old fashioned scientific research. It also shows how the evolution of one adaptation, can benefit or lead to new adaptations. The new head shaped which improved vision, lead to the other senses being improved and the ability to hunt and swim being improved. So, what does the hammerhead do? Well, it does just about everything.

How Psychics work

"Here's something to think about: How come you never see a headline like – PSYCHIC WINS LOTTERY?"

Jay Leno

I am well aware that the world is full of amazing wonders and seemingly remarkable miracles. As a student of science many of these wonders seem all the more remarkable as they take place in a world free from the paranormal. There is nothing in the paranormal world that cannot be easily explained through science or logic. The majority of paranormal phenomenon actually often takes more effort to remain to appear supernatural, than it does to explain rationally. Occam's razor, the simple principle of logic, is usually all that is needed to brush off most paranormal claims. Occam's razor states that "entities must not be multiplied beyond necessity" which basically means that all things being equal, the simplest explanation is the best. A principle in which paranormal advocates and followers blatantly ignore. It's easier for some to believe a person is talking to a dead person than to believe that they are making it all up.

The topic of this article is a shockingly common sight in today's pop-culture. TV mentalists, fortune tellers, psychics and mediums are an increasingly popular attraction in many different forms and in many different guises. They differ in what they claim to be doing, some are plain psychics, some talk to ghosts and others talk to angels or similar entities. Yet the major way in which they are similar is in that these people all tend to believe that they are using paranormal powers, yet they all use the same very normal techniques to achieve results.

Despite the efforts of personalities such as Derren Brown and Penn & Teller, people still choose to believe that these abilities are supernatural powers. It is quite possible that even after having it explained how psychics achieve these results, many will still prefer to believe that psychics are real, and that I am the one who is trying to fool you.

All psychics work using the same techniques, they use a mixture of high probability guessing and cold reading.

Firstly, it is important to realise that most people that attend a psychic reading, want the psychic to be true and to work. They want to know their future and that's completely understandable. The problem is that this leads people to be very easily misled; they will ignore all the misses the psychic makes and get carried away on the hits. For example, my mother was recently at a psychic and was absolutely blown away by her accuracy. The problem she had though was that this lady actually recorded all of her readings onto tape, which my mum brought home. After being fanatically described by my mother how incredible this woman was, I decided to sit down and listen to the tape. While we listened to it, it became clear that the lady actually asked many more questions than she answered and got many more things wrong than she actually got right. By the end of the tape my mum admitted that she had got carried away, and actually it wasn't all that impressive. Most of what she had said was simply wrong, and not even close to any sort of truth. The problem occurs as a result of the way our brains work, we tend to forget bad experiences and remember good ones, which made my mum get very over excited. If she hadn't taped the reading I doubt I could have convinced her otherwise.

The main technique which psychics use is a process known as cold reading. A well-practised cold-reader uses a number of techniques to gain a great deal of information about a person very quickly, often without the subject even realising. They use techniques which closely watch all aspects of the subject, which allow them to make informed guesses about them. By reading body language, race, level of intelligence, the way they talk, the way the dress, the ways in which they respond to certain stimuli and many other things cold readers can pick up a great deal of information about a person. They then use high probability guesses which allow them to pick up more information about the subject by the way they respond to them. They can then tell if they are on the right path by the signals given off by the subject and focus on those areas.

The basic process of cold reading will begin before the actual reading seems to start as the cold reader may indirectly suggest to the subject that they should help them. The cold reader may suggest that many of the psychic images that they see may mean more to the subject than to the psychic, so if anything they mention could mean something to let them know and they can uncover more things about them. This is one the most important parts of a cold-reading as it allows the cold-reader to talk about whatever comes into their head but they don't need to give any meaning to it, as they wait for the subject to find any meaning. So the cold reader actually does nothing but mention a bunch of things which the subject then desperately tries to find meaning in, eventually they will find something. Once they have made a connection, the cold reader will make a series of probing statements or ask questions using special techniques which I will describe later. The subject will then (often unknowingly) reveal further details about themselves or their family or life which will give the cold reader a base from which to work from. Whilst most information seems to be coming from the reader, the reality is that all the facts and statements of actually come from the subject. The cold reader will then sum up and refine the jumble of facts given by the subject in a manner which makes it seem like the cold reader is making a psychic statement. As the majority of the time is focused on the occasional hits the reader has managed to produce, and little time is spent on the much higher number of misses. The impression given that the reader knows much more than they actually do.

Cold reading is a well understood and well-studied branch of performance illusionism. There are over twenty different techniques, which can make the reader seem to perform all sort of illusions from palm reading to tarot cards. The most widespread however are the techniques used to fool people into believing the reader has psychic abilities.

Shot gunning is probably the most popular and widely spread technique used by many TV psychics and mediums. It is so common as it is so easy and affective a way to gain large amount of information from a person, very quickly. The technique involves the reader offering out a large amount of extremely general information, often to a large group of people. The information is off the top of the reader's head and will be almost entirely rubbish, however they will offer up so much information, so quickly and it will all be of such a general content that eventually someone will pick out something that they feel may relate to them or is at least close. Once this closeness is achieved the reader will continue with the technique but will narrow the scope of the information, refining the statements according to the reactions that they receive from the subject.

For example, the reader might say:

"I see an accident with a family figure, a father, uncle, grandfather, cousin (and may continue until the subject responds) ... definitely some sort of accident, possibly involving a leg or arm injury ... possibly an injured leg or ankle"
"I see another person; they are not a blood relative but they are close to you or someone in your family when they were growing up. I can see a darkness in their life, possibly heart disease or lung cancer"

This technique is called shot-gunning, due to the manner in which the reader fires a barrage of small projectiles towards the target in the hope that they will get a few hits. As you can see from the examples I gave above the reader will give a large number of possible options and leave it up to the subject to choose which ones are hits. Once the reader has a few hits it becomes easier to make other hits as the reader quickly gains a great deal of information.

Another technique is known as Barnum statements, and are very simple statements that can be made by the reader which seem extremely personal, but can actually apply to almost anybody. The statements are usually open-ended or give the reader a large amount of room to adapt after they are made. The point of Barnum statements is to allow the reader to gain lots of small amounts of information about a subject, which can later be grouped together and made into seemingly impressive statement. The technique relies on the eagerness of the subject to fill in the spaces and make connections and again takes advantage of the subjects urge for the reader to really be a psychic. Very often the reader will even bully a subject into believing connections that don't really exist and will even accuse the subject of being foolish or forgetting things of importance.
For example, a reader might say;

"I sense that you sometimes feel insecure, especially when meeting new people."

It is very easy to tell a nervous person and most people are nervous around new people, however a subject might feel this is a very accurate and specific reading, when it is really anything but that.
"your father passed on due to problems in the chest or abdomen"
A person who is of a certain age is extremely likely to have a dead father and the vast majority of conditions which are likely to kill an elderly man are located in either the chest or abdomen. So again, this

seems like they have just made a psychic reading but really it is just common sense.

Other techniques include things such as the rainbow ruse, which allow the reader to cover all their bases by offering statements which have both one specific trait and the polar opposite of that trait. This allows the reader to work out which of the two extremes the subject is closer to and make seemingly accurate deductions.

For example:

"I can sense that you are very often the heart and soul of the party, yet at times you can also have moments when you like to be on your own, in peace"

This statement might seem very specific, yet it is easy to tell people that are confident and even if they are not, they would probably like to believe that they are. Yet at the same time everyone has moments when they like to be alone and at peace. The rainbow ruse gives the impression that the reader knows the subject personality extremely well but the reality is that they are just making it up again.

Conclusion
Cold reading is not a secret. Neither is it some sort of special gift that takes huge amounts of experience and practice. It is a well-known technique used by stage performers for over a hundred years. It is not until fairly recently, when the world became obsessed with the paranormal, that some of these illusionists started using these techniques to pretend to be psychic. This was not appreciated much by the world of stage illusionists.

The truth is that it is a con; these so called psychics are taking advantage of the fears and sometimes even the suffering of the general public and using it to make money. Most psychics are not even especially good cold readers, yet they are good enough to fool people who are not familiar with what they are up to and want to believe that they are being told their future or speaking to their dead relatives.

NB. Derren Brown the well-known British illusionist and non-psychic, enjoys re-enacting all of the so-called psychic tricks, before fully explaining them to the audience. Show the audience how they are done without any supernatural powers. Recently he went one better and actually correctly predicted the national lottery result, live on British TV. Something no psychic would ever have the guts, or ability to do. Yet again many people refused to believe he wasn't psychic!

Is God real?

"Is God willing to prevent evil, but not able? Then he is not omnipotent.
Is he able, but not willing? Then he is malevolent.
Is he both able and willing? Then whence cometh evil?
Is he neither able nor willing? Then why call him God?"

Epicurus

Science is described as the effort to discover and to increase human knowledge of the physical world. God is described as a common way of referring to the principle or sole deity in a religion. If we are to accept the concept of God as a reality, is a real thing then he must abide by the same physical laws and constants which he created.

So is god real, or is he merely the childish construct of the ignorant and technologically primitive pseudo-society of Dark Age humanity? Can our modern understanding of the world shed some light on what god is, or do we all just need to grow up?

Before approaching the bigger question concerning god's existence, I believe that we should start with the less important and much easier question of god's appearance. Most people, at least when considering the Western, Abrahamic god, consider the appearance of god to be similar to the famous painting Inside the Sistine Chapel by Michelangelo. He is generally seen as a patriarchal old man figure with a long white beard, long white hair and long white robes. This anthropomorphic view of god is logically speaking, extremely unlikely. Science has proven that humans in their most modern form have existed on earth for around 200 thousand years, whilst the Universe is known to be at least 13.73 billion years old. Therefore, there is no logical reason why the god who created the universe would resemble one particular species which, isn't going to appear on the planet for another 13.7 billion years. If we were looking for a suitable appearance for god, based on the time of appearance and the numerousness on our planet, then god should be a bacterial cell. As the ancestor of bacteria first evolved over 4 billion years ago and for at least 3 billion years nothing but bacterial type life existed on earth. So if god created life in his own form, then he should be a single-celled organism similar to bacteria.

Incidentally if you were to analyse all the life existing on earth, it would actually appear as if bacteria was the dominant life form on our planet, as in each hand full of soil exists more individual bacteria than all the other individual animals alive on the earth. Every other type of life contains bacteria, and everyone is completely reliant on it for its own existence. All other life on earth is nothing more than the bubbles upon the pint of beer that is life. Bacteria makes up the drink and upholds the overwhelming quantity and diversity of life on earth. Although obviously we cannot speak for all potential life on other planets until it is discovered, it is expected that it will be bacterial life that we will discover all throughout the universe. Adding further strength to the argument that god must be a bacterium.

The philosophy of God
Philosophy is based on logic therefore in this particular incidence I believe it to be a useful ally in this argument.

Many religious types feel that the existence of god is a super-natural entity, and therefore is non-empirical and have no place in the field of science. Others, such as Richard Dawkins argue that the existence of god is really an empirical question, as a universe with a god would be notably different from one without, which would be able to be scientifically measured. Put simply, if god exists we should be able to see evidence of his existence all around us, all of the time. I personally believe that this point of view is correct and that if god created the god he would have an effect on it. *

To prove that god existed what we would need would have to be empirical evidence that was scientifically recorded under controlled conditions. No eye-witness reports, no photos, no preachers, no "I see him in the eye of every new born child". We would need cold hard data that is recorded and able to be repeated all over the world in numerous locations by numerous researchers, with the same results. Unfortunately, this empirical data is non-existent which suggest that god to might be that way also.

The Church however believes that the existence of God can be proved by unaided reason. They had to introduce this idea as the freethinkers began using logic and reason to argue against the existence of God. The church felt they needed to end this and therefore their own idea to prove the existence of God also using unaided reason.

The argument of the First Cause suggests that everything we see in this world has a cause. It states that as you go back in the chain of causes further and further you must come to a First Cause, and to that First Cause is God. That argument doesn't carry much weight is modern times as philosophers and scientists have shown this argument to be invalid. There is in fact no reason that the world could not have come isn't existence without cause. If everything must have a cause, then God must have a cause. Basically but, who created god? If we take that angle then we must also think about who created the entity that created, and so on. If as some argue god just appeared from nothing then god has no cause, and if god has always existed then there is also not cause. There is no reason why the universe the needs a cause, it could it not have always existed? Although it is difficult to imagine this, if is possibly just due to the limits of our intellect. Just as a rabbit doesn't have the ability to understand maths.

The argument from design is one of the best know arguments as proof of god. It is the idea that because everything on the planet works seems so well adapted to everything else in the world, that is must have been created this way. For example, the philosopher Voltaire once remarked jokingly, that obviously the nose was designed to be able to fit spectacles. Since Darwin discovered the theory of natural selection, we understand much better why living creatures are adapted to their environment. It is not that their environment was made suit them, but that they adapted through evolve to better fit in to it. There is no evidence of design about it.

Often suggested by followers of this argument is that some things are far too complex to have evolved. For example, what use is half and eye? The simple response being, twice as useful as a quarter of an eye. The eye clearly isn't as complex as we think, as the eye is known to have evolved somewhere between 50 and 100 times. There are examples throughout the animal kingdom, and fossil record of all different stages of eye evolution.

The argument from poor design is a particularly poor one as there are significantly more occurrences in nature that suggest that god was clearly not responsible, by poor design. There are a huge number of poorly designed, suboptimal, confusing and just plain lazy designs out there. Ironically the human eye being one of the best examples. The vertebrates and molluscs both evolved the camera eye separately, the molluscs however did a better job of it then we did. This is slightly confusing because molluscs appeared on earth millions of years before us, why would god give humans, the very species made in his own image, an inferior eye to an octopus? The human eye is inferior that it had to evolve in a fashion that makes it appear inside out. The nerve fibres pass in front of the retina and there is a blind spot where the nerves pass through the retina. This means that the brain has to compensate for the missing piece of information by just plain making up what it sees.

Whether it is useless wings of the ostrich or the 20ft of extra laryngeal nerve in a giraffe or even just the existence of male nipples, the natural world is full of examples of poor and/or sub-optimal design. Every species has many things in its design which could work better if it were designed, individually on a drawing board. Of course they are really a product of evolution, and therefore had to work with what they had, passed down from their ancestors and moulded into the best they could manage. If god laid a finger in evolution it would be blatantly obvious.

Also used is the moral argument which suggests that the existence of God is required in order to bring justice into the world. The argument is a complex one, in that we know there is great injustice in the world, innocents often suffer, and cruel, devious and evil people often prosper. If you are going to have justice in the universe, then there much be something which exists to redress the balance. So it is suggested that there must be a God, as well as a heaven and hell, in order that in justice will be served. This is clearly a fallacy if you look at it from a scientific point of view, Bertrand Russel gave the example that, "Supposing you got a crate of oranges that you opened, and you found all the top layer of oranges bad, you would not argue: "The underneath ones must be good, so as to redress the balance"; this is clearly nonsense.

One of the simpler paradoxes against the existence of god is the omnipotence paradox; "can god create a rock so big that even he can't move it?" Any entity worthy of being considered god must be all powerful, in that there is nothing god cannot do. However, if god cannot create a rock that he cannot lift then he is limited and therefore not all powerful; if he can create a rock that he cannot lift, he is also limited and therefore is not all powerful; therefore, god cannot be all powerful. This argument doesn't deny god's existence however it is useful in showing how the literal interpretation of the biblical god is logically impossible, and as with most discussions about god reduces itself to nonsense.

Problem of Evil
One of the philosophical arguments against god argues, why would the all-knowing, all seeing, all loving, god allow bad things to happen? Evils such as war or disease, criminality, natural disasters etc. it is difficult to imagine a God who gets pleasure from contemplating such tortures. If there were a God, and he were capable of such malevolence then he doesn't deserve to be worshipped. Either that or the exists of evil suggest god doesn't exist.

This argument from non-belief is similar to the aforementioned argument from evil as it reminds people of the differences between the world which does exist, and the world that would exist if god were living in it. The argument from non-belief states that if god truly existed then there wouldn't be any people who didn't believe in him. God would allow or create situations in which his existence could not be denied, was made evident and the people could be given a clear reason to believe. In reality the opposite is the case, there are large numbers of non-believers and no one true faith is established. There is no evidence of god's existence anywhere except in the minds of the believers.

So does god exist?
With the growth of science and knowledge of the ways of nature, the clergy have been fighting a losing battle. They have tried their best however to prevent the rise of geology; they fought and are still fighting against Darwin, they fight against scientific theories over every type and an in all instances come up second best.

I have reached a point in my discussion of the nature of god were I realise that I could argue almost infinitely against god's existence. It would be an extremely one sided argument with the other side maintaining the fact that god's existence is true merely based on circular logic, stating that the bible says god is real and the bible is the word of god, and god is always right and that must be true because it is in the bible etc etc ... ad absurdum. So I don't really see much point in continuing to argue against god.

The reality is that science has no need for god to explain the world we see. We understand the world around us and are continuing to understand it better as each new day. Never do we really have to turn to god for understanding or explanation.

However, billions of people do, and humans as a species seem to have a compulsion to believe in supernatural beings. It is merely based on fear, the fear of the unknown, of and the fear that we are alone and powerless in the universe we don't understand. Science can help us to get over this fear, science can teach us no longer to look round for imaginary supports for our fears. No need to create imaginary saviours in the sky so save us from the unknown. For thousands of years all varieties of places of worship have offered us a place of security, in a modern world it is knowledge which gives our existence meaning

The scourge of the Scottish Summertime

"A midge in your hand is worth two up your kilt"

The above is a well know Scottish proverb about the Scottish midge, and although I am not certain if there is some sort of a hidden connotation or innuendo to this phrase, it certainly makes a lot of sense if taken literally. I'm sure everyone that has been to Scotland in summer(ish) time, will have experienced the microscopic menaces, as they gorge upon their bare skin. The Scottish Midge is a famously intolerable insect *rsehole.

I feel that the midge almost epitomises the Scottish summer both literally and metaphorically. Metaphorically speaking, the midge sums up what it is like to live in Scotland. In my personally opinion Scotland it one of the most beautiful and magnificent countries in the entire world, unfortunately spoiled by a few minor (and not so minor) problems. The midge mimics this by so often ruining our otherwise very pleasant days. We wait so long for nice weather, put up with so much rain and dullness, and when we finally do get some nice warm weather ... the midges appear to ruin our good times. It's just typical isn't it?
Being a person who apparently has very tasty blood, I have often wondered (usually in an arm-flapping rage) what the point of them was. Well the truth of the matter is, like all things, that they don't really have a point!* Their main purpose of existence is simply to create more midges. Which is fair enough, as the same can be a said of all organisms. A chicken is just an eggs way of making another egg. Things don't need to have a point, and they certainly don't need to be of any use to US to exist. Evolution or more specifically DNA, made the midge, just like it made the human being, simply as a vehicle to continue to exist!

Well if you have ever let a midge sit on you for long enough before squashing it (which would be a bit foolish to be honest) you will know that the midge is a very small type of biting fly. To be fair, not all types of midge actually bite, but the ones which we care about in this chapter do bite (a lot.) They are known as *Culicoides impunctatus* or the Highland midge. They are generally considered to be the most annoying type of midge, which is nice, but I'm thankful I don't live in the highlands where they are apparently even worse. In one study conducted in the north-west of Scotland, five million midges were collected from a two metre area of skin. In the highlands they have to go out wearing mosquito nets, midge repellent is not affective enough.

Although you might believe midges number one purpose on earth was to annoy you, like all live the adult midge's main priority is sexual. The male midge spends its time looking for a partner, it is the female (concerned with feeding its young) that will bite you. In fact, the female cannot lay its eggs until it has had a blood meal. After which it lays its eggs in a semi-aquatic habitat, usually in some sort of damp crevice such us under tree bark, in compost or some other type is similarly unpleasant location. So every time you kill a midge you are stopping it breeding so... good job!

It is in practice, but not literally impossible to avoid being bitten by midge. In a literal sense, if you know the midge well enough it may be possible to avoid them, but practically speaking it is not going to happen. However, knowing what midges like and what they don't like will help you to avoid them slightly, or at least understand why you are standing in enormous clouds of the buggers.

Midges like ...

- **damp conditions**
Standing admiring Scotland's beautiful lochs and fly-fishing in our famous rivers, are often ruined by their mutual attraction to these little biters.
- **still air**
Preferred by the midges as it makes flying easier, luckily a rare thing in Scotland
- **warm weather**
Insects are cold blooded so help the midges to reproduce and to digest their blood meal, luckily also quite rare in Scotland for most of the year.
- **the shelter of trees**
The trees help the midges to keep away from the unpredictable Scottish weather conditions
- **mornings and evenings**
Because they know we mostly work 9 to 5
- **dark clothing**
Because they know black is slimming. No, it probably for camouflage reasons.

\- shade

To relax.

Midges DON'T like ...

\- dry conditions

The weather is very rarely dry, especially were midges are common. Scotland has a temperate climate which means it has fairly high humidity most of the time.

\- wind

Midges cannot fly or detect chemical trails as affectively in high winds so don't enjoy it.

\- cold weather

Easy to understand, nobody likes the cold.

\- bright and strong sunlight

I'm not sure why, maybe it's the blood-sucking/vampire connection.

\- exposed areas

They won't fly too far from tree, as I explained earlier

\- light clothes

Not certain, may be repelled from light colours as the think is actual light or because they are dark and they stand out against the light background.

\- Insect repellent

It also may be important to know when trying to avoid midges that they are attracted by the carbon dioxide, which we release when we breath. So avoiding large crowds would be another "midge-evasion" technique with potential. Finally, you may have noticed that when you first notice the midges have arrived, their numbers are usually low but steadily increase as time goes on. This is due to the unfortunate fact that midges release a chemical whilst feasting on blood which other midges have developed the ability to detect, giving the impression that they are inviting their pals along to join the party.

In conclusion

There can't be much said for midge to be honest. They are annoying, they take advantage of us and they leave us with bright red welts many times the size of their own bodies, which itch for days. However, although I doubt too many people would miss them if they were to suddenly vanish (and global warming may oblige) I believe that the midge is as much a part of Scotland as the haggis, the kilt and disliking the English (joking). Deep down we all harbour a little affection for our minuscule monster mates and Scotland would be a lesser place without them*. I'm not sure how, but that's the way I choose to end.

Why you are always in the bathroom when the postman calls?

"Coincidence is the word we use when we can't see the levers and pulleys."

Emma Bull

It's one of the many mysteries of life, why, oh why, when you have been waiting all day for an important package to be delivered, does it always get delivered at the one time you don't want it to be?

Imagine the scenario; I'm certain it's a familiar one to most if not all of you. You are waiting on a package, its important so you have been deliberately putting off going shopping, going to the gym, walking the dog etc, etc. You have to do this because you only have one shot at receiving delivery, if you miss the delivery you will have to go pick it up at the post office headquarters, or even worse, it might be returned to the sender.

You need to go to the bathroom, you figure out that if the postman hasn't come in the last 8 hours you have been awake, what are the chances that he will come in the 5 minutes you aren't able to answer? So you go, but as soon as you sit down … ding, dong!

What are the chances? It may seem hard to believe that seemingly random event appearing to occur more often than logic would dictate. This is because the world is so full of coincidences but there are lots of examples of them occurring all around you every day. It is hard to understand how such a random, yet specific part of your daily routine might continuously become the victim of such a chance occurrence. It sounds completely ridiculous to even suggest that there might even be anything even vaguely scientific in it?

You might suggest that it was just bad luck, something like Murphy's Law and coincidence mixed together, and you would be correct in a manner of speaking. Be that as it may, there is actually an entire branch of mathematics named after Frank P Ramsey, known as Ramsey Theory which attempts to explain just this sort of occurrence.

Ramsey Theory attempts to explain the conditions under which order must appear, and tries to make sense of complete nonsensical events such as the postman always arriving when you are in the bathroom. Ramsey theory states that a sufficiently large system, no matter how random, must contain highly organized subsystems. Simply put, it tries to understand and simplify the conditions under which order must appear and states that complete disorder is impossible, even out of total chaos. So what does this mean to the real world? Well if we take a few steps of logic, it means that no matter how much of a coincidence or how random something might seem, mathematically speaking it was neither and it was inevitably going to happen. It goes some way to explain why the world is so full of coincidences and chance occurrences, which I will now try to explain without using maths. Thank goodness.

"*Every single moment is a coincidence*" Doug Coupland

Doug Coupland is a Canadian novelist and he isn't really relevant to this chapter other than he does seem to have the right idea about coincidence. Coincidences are happening all of the time; every single minute of every day. We shouldn't even still be surprised by coincidences, as so they occur so often. Even still, the dullest and simple coincidences can still quite fascinating phenomena when you look at them closely.

For example, I missed the train home by a mere 30 seconds a few days ago after meeting a friend in town for coffee. Now, there were an almost limitless number of things that I could have done before, during and after stopping for a coffee that would have culminated in me not missing the train. For example, I might have not decided to go the bookshop beforehand, as it had me waiting for 2 minutes; I might have brought cash so I didn't have use my credit card to pay for me as well, I might have walked to earlier train, so I would have known what time my train was at. My point is that there were hundreds of instances where I lost or had possibly even gained seconds, which eventually lead to me missing the train. All of them coincidental, all of them random and all of them culminating in an event which set up yet another chain of events, which could not have occurred without the earlier events occurring. If I hadn't missed my train, I wouldn't have met another friend on the walk between the train station and my house, and got a lift home in his car, which was nice because it was raining.

In one of my favourite books of all time, 'Unweaving the Rainbow' by Richard Dawkins, the author goes quite deeply into the subject of coincidence. In his book he coined a new term "Petwhac" which is extremely significant in our interpretation of Ramsey Theory, as the two appear to be almost one in the same. PETWHAC stands for a Population of Events That Would Have Appeared Coincidental. It refers to incidences which appear to be coincidence but when looked at more closely, are actually both very possible and statistically probable. The example that Dawkins uses in his book is of a television psychic who promises to stop peoples watches, through the TV using only the power of their mind. The psychic asks anyone whose watch stops to call in and let them know. First it can be calculated that the probability of any given watch stopping within any given a five-minute time period is around 1 in 100,000. Not that likely eh? However, if you consider that there are roughly 10 million people watching the show, and assume that half wear watches, we could statistically expect that about 25 watches would stop. If only a quarter actually called into the show that's still 6 people whose watches stopped, apparently with the power of the psychic mind! Add to that the people whose wall clock, alarm clock, any type of clock had stopped. So, an event which appears to be hugely coincidental, so much so that it seems paranormal explanation would be the only explanation, turns out to be neither.

I have experienced myself similar seemingly unexplainable feats of outstanding coincidence, which at the time seemed baffling but when broken down seem almost inevitable. Recently I was driving toward Glasgow, leaving my home in Greenock and I passed a 4x4. I recognised the car as it had a spare wheel cover which had the emblem of my old rugby team on it. On the way home, about 3hrs later, I overtook the car again, this time coming into Greenock. Seems strange, but not if you consider that we were both driving to watch the biggest rugby match of the year, which was in Glasgow. It started at the 7pm, so we would leave Greenock at the same time and it ended at the same time some we wound leave Glasgow at the same time, reaching home at similar time. I drive fast, so if he was in front of me I was more likely going to catch up with him in his mush slower car. I was certainly going to spot the car as it was so instantly recognisable to me. It seems like a one in a million chance, but again it was really quite likely*

Conclusion
So, when was then last time the postman called when you were in the bathroom, can you remember such an occurrence or something similar.

Coincidences are an extremely common part of most people's life and I'm sure you can all think of times when Ramsey's law was all too real in your life. An example of this occurred just a few days ago, when my neighbour came over with her entire family (including two children) so that I may show them my pet snakes while I was in the middle of enjoying a bubble bath. You have no idea how hard it is to handle a 7ft Boa Constrictor while trying NOT to drop your towel. Not massive coincidence but I was caught in the bathroom.

And for the record, I was eventually able to avenge this coincidental embarrassment of occurrences by intercepting a postman's delivery that I was supposed to miss. I was expecting a parcel delivery some time over the next 4 days, but I had to go to the gym, so I took a chance. As I started walking up the road I realised I had forgotten my gym card, and I had to go back to my flat. I found my card and as I was leaving a second time, a postal van pulled up outside my drive with my delivery. The world is full of coincidences but they are not to be feared as order is always restored...even through complete chaos (or in my case forgetfulness)!

*just in case you have completely lost faith in coincidence, I went to put batteries in an old clock the other week and when it started working again I realised it was at the exact correct time. OK, it was out by 3 minutes but it was almost exactly correct!

What is ringing in your ears?

"This morning I can hear the roar
Of whistles deep inside my craw,
The hiss of steam is most insistent,
I hope today I'll prove resistant"

Tinnitus C by Mr Gerald B Frank

We have all woken up with ringing in our ears I am certain, as a fan of noisy bars, nightclubs and rock music I can particularly empathise with Mr Gerald B Frank. He actually wrote a series of at least seven separate poems about ringing ear noise, so assume he suffered quite a bit. This man likes to write poems about what he is going through and like him, many people like often complain about ringing ears. But, how many people actually know why their ears are ringing? Obviously it can be caused by excessive exposure to loud noises the previous night, and the damage caused to your ears. However, what is it about the powerful sound waves that has caused physical damaged and why do you now here a ringing and where does THAT noise come from?

The condition which causes the ringing in your ears is known as Tinnitus. It can be experienced with an array of hugely varied sounds. From buzzing to humming, whining to hissing, ticking to clicking, whistling to screaming and whooshing to roaring...all the audible bases are covered.

The Ebers Papyrus, which dates to 1550 BC and is one of the most important medical documents of the ancient Egyptians, is said to have made a comment referring ringing in the ears. Although the validity of the claim is controversial, the papyrus is often claimed to be the earliest historical reference of tinnitus. It is rather likely that tinnitus was a known symptom in the ancient Egypt but it has never fully been confirmed to have been treated.

Tinnitus is not, as I might be making it sound, an illness or disease but is a symptom. That means that is just a feeling, noticed by a person which might indicate to an abnormality. There are a large number of causes which can result in tinnitus. The most common examples include the build-up of wax in the ear, ear injury or getting something stuck in your ear. Usually ear injuries happen either from the intrusion of a foreign body (finger, pencils etc) or from a loud noise (scream, music etc). That is why you are always told not to use cotton buds to clean your ears and not to turn the volume on your earphones too high.

Obviously the most common cause of tinnitus, temporary and permanent is exposure to loud noise. Whether it is a job working as a DJ or from using a vacuum cleaner all day, all load noises are at serious risk and preventative measures should always be used if possible. Tinnitus can also be caused by head injuries such as whiplash injuries were the head is jerked violently, or head and facial fractures.

Other causes of tinnitus include things such as ear infections or ear wax impaction. Some medications can cause ringing in the ears, even some over the counter drugs such as aspirin have been linked to temporary tinnitus. Other medications that you might receive from your doctor, such as several antibiotic drugs have also been linked to tinnitus, as have several anti-viral, anti-cancer and anti-depressive drugs. Tinnitus is also more common in people who have psychiatric problems such as depression and anxiety. I can imagine ringing in the ears is the least of their worries but a drug that magnifies it can hardly help the situation.

The list of possible treatments for tinnitus vary from the bizarre the absurd. For example, "Transcutaneous electrical nerve stimulation" or "Direct stimulation of auditory cortex by implanted electrodes" or even taking Ginkgo Bilbao. The main thing as these treatments have in common is that they are all nonsense. Although there are many treatments claimed for tinnitus there are none that have had any success.

I you do suffer from tinnitus; you are not unique as it is very common. According to wikkipedia.org famous people with recurrent tinnitus include; Sir Richard Attenborough, Ludwig Van Beethoven, Eric Clapton and Vincent Van Gogh, so you're in good company.

The Common Cold

"Give ear, you scientific fossil!
Here is the genuine Cold Colossal;
The Cold of which researchers dream,
The Perfect Cold, the Cold Supreme.
This honoured system humbly holds
The Super-cold to end all colds;
The Cold Crusading for Democracy;
The Führer of the Streptococcracy"

From "Common Cold" by Ogden Nash

The common cold or *Acute Viral Nasopharyngitis*, is a highly common and highly contagious upper respiratory tract infection caused most commonly by a variation of a picornaviruses or cornaviruses. Symptoms of the cold are usually mild but can be extremely unpleasant and include sore throat, runny nose, blocked nose, cough, muscle aches, tiredness, headaches and loss of appetite.

The cold can be spread from person to person through two mechanisms, through the air, by sneezing and coughing and secondly contact with saliva or nasal excretions from infected persons.

Currently there is no cure for the cold and there is no vaccine. However, it is not possible to catch the same strain of cold virus twice as immunity is built up very effectively by the immune system. This means that once you have recovered from the cold you will not catch it again, so long as a new strain is not introduced. Unfortunately, there are many, many different strains of cold virus.

Against common belief, or what your parents might tell you, you do not catch the cold from cold weather. Getting cold does not increase your chances of catching the cold, and wrapping up warm does not reduce your chances of catching it. There has actually been a great deal of research into the subject, and there has never been anything found linking the cold or cold weather to catching the cold. The cold virus is just as likely to be caught in the middle of summer than in the winter. The cold is more common in the winter however, and the reason for this is indeed linked to the weather, but not in the manner you might expect. The weather is colder and wetter in the winter months and people spend much more time indoors. When people remain indoors more often they expose themselves to a much higher threat of coming into contact with the virus through cross-contamination with an infected person. In a sense, the viruses become isolated and contained in the perfect conditions for them to be transmitted and spread.

Again contrary to what common belief, or your pharmacist might tell you, there is nothing you can buy that will reduce your chances of catching the cold or reduce your symptoms. The most popular ones you might hear from your pharmacist just now are high doses of Vitamin C or Echinacea drops. Both of these have been extensively researched and have yet to find any positive results. Pharmacists are in the business of making money, and are not scientists, so just because they suggest it doesn't make it true. The best solution is to take plenty of rest, take plenty of fluids and the virus will leave your body within a few days. The only thing that is going to get rid of your cold is time.

What is Cellulite?

"Like a swift migrating fish the word cellulite has suddenly crossed the Atlantic"

late 1960's publication of Vogue magazine
(First know reference to cellulite)

If you were to ask a random group of women from the public and ask them what would you most like to change about your body, a significant percentage would mention cellulite. The beauty industry is well aware of this fact. As one of the most parasitic industries on planet, it is based upon exploiting people insecurities and fears. The beauty industry has a lot to answer for, few come close to cosmetics producers over the years when it comes to selling quackery and straight up lying to people.

An example of that is the cellulite was actually first coined by the cosmetic industry, aimed to confuse and frighten the general public. This is one of the reasons that there is never very much said about what cellulite actually is. Cellulite is really just fat stored just underneath the skin, which has accumulated to the extent that it pushes against the strands of connective tissue.

The amount of cellulite that a person has on their body is influenced by several key factors. Your genes, your gender, the amount of fat your body is storing, your age and the thickness of your skin. All affect how much cellulite you have, and more importantly how visible it is. As we are often told, there is actually nothing special about the type of fat in cellulite. It does not contain any unique toxins or chemicals which are stored in it is, and it is not any different than any other type of fat. The reason cellulite seems worse is simply due to the fact that it tends to be visible in areas that are least effective at consuming fat deposits. Therefore, it sits there for all to see and takes a lot greater effort and dietary sense to reduce.

The most important point to take from this brief chapter is that no fancy exercise machine, flashy cream or other quackery is going to reduce your cellulite. There are not any miracle products, treatments or medicines that can destroy cellulite. There are some salon treatments available that reduce the appearance of cellulite. All these treatments however only temporarily give the illusion that the cellulite has reduced and therefore are not really doing anything. There are also expensive treatments such as liposuction and a cellulite destroying drug injection called mesotherapy, but even these are not a cure.

The only sure-fire way to get rid of cellulite, as with all fat in your body, is to consume fewer calories, less fat and to exercise more. Diet and fitness experts agree that the best exercise regime is something that combines aerobic exercise and strength training. That way you will reduce fat and increase muscle tone, which will have the combined effects of reducing flab and firming up.

A permanent change in diet is also required if you wish for cellulite to never come back. Fad diets do not work, they only work whilst you are on them and stop working as soon as you come off them. In many cases you will in fact put on much more weight, as soon as you come off the diet as being on the diet has altered your metabolism, meaning more fat is stored.

If you want a new body, you need a new lifestyle. If your old lifestyle could accomplish the body you wish you had, you would have it already.

Everyone has cellulite, men get it too. The reason it is appears less common in men is only down to the way than men store fat. Men tend to store it in the belly and women in the legs. Cellulite is totally natural and even the thinnest people have small amounts of it. If however, it really makes you feel self-conscious and wish to conceal you it, try using a self-tanner or going for sun beds, as cellulite is less obvious on darker skin tones.

The Science of Bruises

"Bruises linger longer as we age, staking their claim
on the soon-to-be-rotten sweet fruit of our flesh.
A baby's blank skin never holds a grudge for long.
It's a time-lapse miracle, putrefaction in reverse.
The hickeys I left on her neck were collapsed rainbows,
unfolding their colours backwards under her sweater.
The edgy world keeps touching us, to prove it does
exist:" –

Bruises, by Michael McFee

Have you ever wondered why bruises come in so many colours and why they even seem to change colour from day to day? And why they sometimes take days to appear, when surely the damage is done at the time of injury. Sometimes bruises are sore, and sometimes you can't feel a thing! You wouldn't think bruises would be a great topic for a chapter, but they are more interesting than you might think.

A bruise happens when small blood vessels underneath the skin are broken. The reason some bruises take a while to show themselves is that sometimes it can take a while for the blood which leaks out of the vessel to reach the surface. This happens if the damage occurs deep in the muscle tissue, the blood has to make Its way through layers of muscle and planes of fibrous tissue. This fibrous tissue also explains why bruises occur some distance from the original injury, as blood makes its way to the surface wherever it is able to.

The classic bruise colour is a reddish-purplish hue. This colour is created by the haemoglobin in the blood being leaked by the wound. To repair the damage at the site of the injury the body then brings in white blood cells, which causes the red blood cells to break down. This creates substances that are responsible for the first colour change, which is a fading of the red hue.

One of the products of the breakdown of the haemoglobin is called Biliverdin, which has a green hue to it, and creates the green parts of the bruise. The yellowish colour of the bruise is caused by another breakdown product of the haemoglobin called Bilirubin. Which is usually secreted from bile, and is also the substance that gives people with Jaundice, the yellow colour to their skin.

The final stage is that the Bilirubin is taken up into the liver, where it is converted to bile and used in food digestion (as well as giving faeces their colour). At the same time more white blood cells cause more colour fading until the bruise has vanished.

Great, eh?

What is the point in wasps?

I don't like wasps,
so I trap them in pots.
They are very scary,
and even a bit hairy.
Which makes them even more scary.
Oh how I hate wasps.
A LOT!

This is a poem that I found on the internet written by someone called Jenny, and I think it sums up most people's attitude towards wasps quite well. The wasp gets a hard wrap, nobody likes them and nobody has any sympathy towards killing them. They seem to exist merely to torment us, so what is the point in wasps? I mean bees make us honey but wasps, they just annoy us and sting us."

So what is the point in wasps? Firstly, why should they need to have a point that is a benefit to us? There is no reason that any organism should need to help us, the ones which do are merely created or manipulated by us. They did not evolve to help us (see "argument from design")

There are almost infinite species of wasp with extremely different characteristics and similarly infinite lifestyle types. Wasps are extremely important and beneficial to humankind on many levels.

As you have probably assumed, wasps are quite closely related to bee, as they are a member of the same Order of insect (Hymenoptera.) However, they are actually more closely related to ants than to bees. Ants are also Hymenopterans, and some species which are commonly called ants are technically a wingless type of wasp, such as velvet ants, also known as "cow killers".

Wasps are members of the suborder Apocrita, and differ morphologically from the other member of the hymenoptera by having a broader connection between two important abdominal segments. However, the easiest and less boring way to tell the difference between bees, wasps and ants is usually: if it is brownish in colour and crawling on the ground with no wings, chances are it's a ant; if it is fluffy and interested in flowers, it's a bee and if it is mean looking and interested in you, it's a wasp.

That just said, amongst the species of wasps that we are probably most familiar with, there is much less room for misidentification. Either the Common or European wasp will be the most likely to have upset you at one time or another, and is quite hard to mistake either for anything other than a wasp, but very easy to mistake each for one another. You would have to get far too close for comfort to be able to tell the difference and even then, would you really care? Anyway, these particular wasps are members of the family Vespidae, which actually contains around 5000 species of mainly solitary (non-social) wasp species. However, both the Common and European wasp are actually eusocial, almost diametrically opposed to their reclusive family.

Eusociality is a Greek term for the highest level of social organisation and it used by entomologists to describe the social structures of many members of the hymenopterans. In bees, wasps and ants it means that they must have a division of labour which is controlled by the reproduction of related individuals. They must also have overlapping generations of offspring that are being produced at all times. Lastly they must have co-operative care of the newly born offspring. Both species live in nests which are built from chewed up bits of plants mixed with the saliva of the wasps and constructed as a large mass of small cell structures. They are easily differentiated from bee's nests as they are generally smaller, more delicate in appearance. They also differ from hornet nests which are usually up in trees or bushes.

There are so many species of wasp that I'm not even going to mention how important all wasp types really are, as it would take far too long, but bare this in mind. There is an entire family of wasps which entire life cycle is involved in the pollination of figs. Every fig has a number of wasps living inside it and both the fig and the wasp and entirely reliant on each other of the survival of each other. So there is one thing that humans need wasps for. I don't like figs very much, but if you do I hope I haven't put you off them. It is generally the inedible part of the plant that the wasp is found and there is no risk of you consuming a wasp when eating a fig. So if it weren't for wasps there would be no figs, that's a starts at least.

We hate wasps because we think they are mean, mostly because they sting us and don't have the good grace to die after. The reality however is that most species of wasps are softies, at least as far as their diet is concerned. Most wasp species, when adults eat nothing but nectar or fruit. In both the European and Common wasp (as with many others) the adults predate on other insect but they feed them to their young. This of course means that wasps have a very important role in the control of many human pest species such as caterpillars (which are their favourite) and beetle larvae. Without the tireless work of the wasp and many other types of insect, pests would be uncontrollable and causing millions of pound more in losses every year.

They also have another important role in the ecosystem as all predators do, by exposing the weak and allowing the stronger to survive. They assure the survival of the fittest, and by consuming the dead they may also reduce the chances of deaths by infection and disease. So wasps are not all bad, are they? There are probably many other ways in which wasps inadvertently help human beings in various inconceivable ways. I am going to resist however as I don't see it as important, why should wasps help humans?

So, what is the answer then?
What I believe the wasp really exposes is the attitudes that people generally have about how animals must have a beneficial purpose to us. You could blame the bible; you could blame the domestication of livestock but I tend to think it's just human nature. People have an almost irresistible compulsion toward anthropomorphism, it takes an active effort to resist it. We find it hard to remember that Humans are only one of millions of species of life on this planet. We consider ourselves as especially important, all species are probably concerned with their selves above others. As we the only species to evolve intelligence to anything near our level, we consider ourselves more important. We can't fly like a bird or a bat; we can't produce electricity like an electric eel; we can't hold our breath underwater like a Sperm whale and we can't detect electric fields like a shark. We need to make a machine to do all of these for us.

We are a unique species of ape and we have a sort of intelligence which is extremely special on this planet. However, we are not the pinnacle of evolution, it literally impossible for any species, and we are not greater or lesser than any other extant species. We are all at exactly the same point in evolution as every other species; we share a common ancestor with every living species on the planet today. In these terms at least wasps are our equal. So now you can see, a wasp owes us nothing.

How accurate is Jurassic Park?

"Dinosaurs and man... two species separated by 65 million years of evolution, have suddenly been thrown into the mix together. How can we possibly have the slightest idea of what to expect?"

Dr Alan Grant (in Jurassic Park)

When I was a blonde haired, blue eyed ankle-biter, my twin brother and I were obsessed by dinosaurs. We had dinosaur posters, toys, books, replica fossils and even had a dinosaur themed birthday party one year, with a stegosaurus birthday cake that my Grandma Pat made. A decade later, the Hollywood blockbuster that was Steven Spielberg's Jurassic Park, rekindled my love of the most famous sub-order Saurian Reptile known to man.

As a dino-obsessed child from just about as early as I can remember I wanted to be a Palaeontologist, until I realized that it wasn't quite as glamorous as it seemed. To quote a palaeontologist from The Simpsons in the episode, " 'Scuse Me While I Miss the Sky" when Lisa tries to find her calling in life, "don't do it little girl, I've spent 35 years of my life brushing the teeth of dead monsters."

When I was a child my favourite dinosaur was Dienonychus, which was a larger relative of Velociraptor. You can imagine how excited I was when I heard about the Jurassic Park movie, and I certainly wasn't disappointed when the monsters of my dreams came alive on the cinema screen in front of me.

Jurassic Park was originally a bestselling book and airport departure lounge favorite by Michael Crichton, long before the movies version was ever conceived. It is however the movie that is the more familiar to most people, as it was to become one of the most popular movies of all time.

Looking back as a Scientist, if I'm going to be a stickler for detail (of course I am) then is Jurassic Park really as scientifically based a movie as it pretends to be? Or is the entire story just as unlikely as it seems at a first glance? I going to take a brief look at some of the scientific details of Jurassic Park and discover whether or not it is as true to reality as we were lead to believe.

First of all, let's look at the pillar stone of the entire story, when the small amount of dinosaur DNA is removed from a mosquito which is preserved in amber, and the extracted polymerase chain reaction (PCR) is then amplified for use. There are actually a number of problems with this approach from the start.

Obvious mistakes such as the fact that the site in which they collect the amber, in the Dominican Republic, in reality does not have fossil sites old enough to contain dinosaur aged Amber. The Amber found at these sites have been dated to as far back as 30 million years ago, but as the dinosaurs died out 65 million years ago it is not possible that the particular amber mentioned in the film and book had any dinosaur DNA in it. Furthermore, none of the dinosaurs that were in Jurassic Park have ever lived in the Dominican Republic, which is the only place mentioned as a collection site for Amber.

I realize that now that I may be getting a little pedantic but in the movie, the close up of the mosquito shows the distinctive furry antennae of a male. Yet only females enjoy blood meals, so you would not be able to extract any blood from a male except from the blood in its own circulatory system. These could be excused as just incidental detail and don't really affect whether there might be any real science behind the story. More sort of mistakes that a Star Trek fan might bring up at a sci-fi convention, or so popular culture has led me to believe*

There are still several Inconsistencies however. For example, to be able to extract just one dinosaur's DNA it would imply that the mosquito only had a diet of one animal, which is extremely unlikely. It would also be impossible to avoid contamination inside the mosquito which would involve the mixing of mosquito, bacterial and viral DNA, and it would remain mixed together even after the sample is extracted. PCR is also not the best candidate for use in the cloning process as it is very specific, and is also extremely easily contaminated and would therefore be useless for cloning. Mosquito blood is also not ideal as mosquitoes digest their blood meals very quickly, so any blood found in the stomach would be severely broken down, unless preserved whilst in the act of feeding. The process of preservation is also very important; when living things are preserved or fossilized the molecules are altered and would not necessarily remain intact. Amber is probably the best place to try and start however, as it can preserve organic matter, although still extremely unlikely, even impossible to be able to preserve DNA for tens of millions of years. DNA, no matter how well preserved can't even last a million years, and sap-producing trees didn't even exist in dinosaur times. So as you can see the procedure mentioned in Jurassic Park would just not be possible.

Another important part of the movie comes about when it is mentioned that all of the dinosaurs are bred to be female to avoid reproduction, stating that all dinosaurs are female by a sort of default setting. The film then goes ahead to try and explains that all vertebrates are inherently female, and need the addition of another hormone to turn them male. This is not actually, vertebrate embryos do have a sort of default setting, but it is neither male nor female. it would actually be an undifferentiated cell, possessing organs which can turn either male or female. Dinosaurs are actually most likely to be the other way around, as in both birds and reptiles (the closest living relatives of dinosaurs) as the chromosomes of birds and reptiles are the opposite way around from other vertebrates, with males possessing two matching chromosomes instead of females. This would have made the dinosaurs on Jurassic Park either normal males or sterile males with an extra chromosome, but certainly not females.

The Dinosaurs
Let's just imagine for a second that everything I just mentioned was not true. They had found a good source of fully-intact DNA from the correct Epoch, have overcome all the biotechnological hurdles and have managed to produce living, breathing, snorting, terrible lizard. Did they manage to get the dinosaurs in Jurassic Park correct?

Procompsognathus
Better known as the "compy's" which are delightful cute little mini monsters. In the book they are portrayed as being venomous, certainly not a fact that is justified anywhere in the fossil record. Crichton also has the compy's acting as scavengers and coprophagists (poo-eaters) as their main role was to clean up the waste of other dinosaurs. He clearly had not done his research as it what would a scavenger need venom for? Unless it was not a true venomous bite, but like the bite of a komodo dragon was mainly derived from the bacterial fauna of its mouth. The fossil evidence proves the compy had very small fine and numerous teeth and long narrow snout and jaw, the exact opposite of most scavenging animals.

The film was a little more accurate as it had the Procompsognathus acting as tiny fleet-footed predators with long tails and sharp claws and teeth. There is no evidence that they lived in large groups but it is not altogether impossible. Just like present day small predators like Meerkats, they would have been an easy target for many larger predators and would have greatly benefited from any sort of group structure. So far the dinosaurs are holding together slightly better than the DNA.

Velociraptors

The "raptors" were one of the stars of the films and the book so you would expect that every effort would be made to make them as accurate as possible. Unfortunately, this was not the case. The raptors in both the book and the film were actually much larger than any living Velociraptor ever was. In Jurassic Park the idea was that the raptors would be a roughly human-sized animal, however the true size of Velociraptor was more closely comparable to that of a turkey. Not quite as threatening an idea to have a pack of scaly, super-smart turkeys chasing you through the forest. Incidentally there did exist a species of raptor that is roughly the same size as the film raptors, known as Deinonychus (as I mentioned earlier). Deinonychus was slightly smaller than the raptor in the film, about 3m long and 75kg, it might have been able to eyeball a small human if it stood bolt upright. Other than the few inch in height, it is a perfect fit for the raptor in the film and exactly why this name was not used in the book and film is a bit of a mystery, maybe it isn't scary sounding enough a name.

Interestingly there were also raptors that were much larger than Deinonychus, such as Utahraptor, which is known to have grown to over 7m long (possibly up to 11m), and weighed at least half a ton. Did Spielberg perhaps consider these raptors too formidable for Jurassic Park, I guess you need to have at least a whisper of hope to allow our human's heroes to be able to escape.

Another reality of the Velociraptor which would have possibly made it appear less scary in the movie is the addition of feathers. Although it was not confirmed at the time of the book, it is certain now that all raptors had feathers. They are in fact the direct descendants of all living birds. Feathers are really little more than highly modified scales, and large numbers of dinosaurs have been found to have had feathers. In the third film, the male raptors were given a small feathered crest; but feathers were actually found covering the entire body of Velociraptor and its kin, including long arm feathers giving the illusion of tiny wings and elaborate tail feathers. Velociraptors would have indeed looked very much like a killer turkey. Some have even suggested the T-Rex would have also had primitive feathers to help keep it massive body warm.

The final thing that made the raptors so frightening and dangerous was their intelligence. In the book and film the raptors had a level of intelligence almost on a par with humans, and were able to solve complex problems and come up with difficult plans, even working out how a door works. Although the raptors were certainly one of the most intelligent dinosaurs, this is really a relative term and it is unlikely that they were anywhere near as intelligent as in the movie. By taking a plaster cast of the braincase of Velociraptor it has been possible to work out that did not actually have a big enough or more importantly, the correctly shaped brain to handle the complex thoughts and tasks featured in the book and film. Raptors were certainly very smart dinosaurs but that's sort of like saying a big spider, it might be big for a spider but you can still step on it.

Dilophosaurus
If you thought that the raptors were bad, Dilophosaurus was probably the least accurate of all the dinosaurs in Jurassic Park. In the film the Dilophosaurus is a 4-foot-tall, skinny, multi-colored creature with a massive frill around its neck and a venomous bite and spit. In the book however there is no frill and the animal is a lot larger. The truth about Dilophosaurus is that it is actually a relative of tyrannosaurus, be that a small one, but it is still a relatively bulky and large (20ft long) dinosaur but otherwise pretty dull animal. When I say dull what I mean is that it doesn't really have a frill around its neck or spit venom, it is a pretty impressive predatory dinosaur in its own right, though almost certainly everything that you liked about that animal was just made up. In both the book and film the animal spits venom and has a venomous bite however there is no fossil evidence to suggest Dilophosaurus was venomous or had a neck-frill. It did have two large bony crests on the top of its skull and that is actually what Dilophosaurus means "two-crested lizard". It is thought that the crests were used for some sort of a display, possibly either a threat display to scare of rivals, or a display of health to avoid fighting. This display would have been useful to Dilophosaurs as by looking at the tooth and skull structure it has been suggested that they were scavengers. Therefore, a device for

minimizing confrontations between rivals for food would have been a great benefit for reducing injuries.

So again the real dinosaur was slightly different from the book version of the beast, and nothing like the film version. One thing that didn't understand about this animal why Dennis Nedry let that little Dilophosaurus kill him so easily. All he had to do was get out of the car and scare it away, he would have been twice the size of it and it would have run away as soon as it had seen him stand up.

Tyrannosaurus

Most people's favourite dinosaur, and for good reason as it is a very impressive creature. Up to nearly 45 feet in length and 7 tons in weight the T-rex is a truly colossal beast. The most impressive part of the T-rex however is it huge head. The skull is to 5 feet in length, the morphology suggests the skull evolved to be a light as possible, while retaining the ability for the enormous muscle attachments and a well-developed brain.

However, does it really lose sight of a person if they stop moving, as suggested in the movie? To be fair to the T-Rex, that's not really even an important question because even if it couldn't see you it would certainly be able to smell you, as scans of the brain cavity have shown a highly developed sense of smell. These same scans as well as numerous other tests have shown the reality is that T-Rex actually has extremely good binocular vision with a high visual acuity, and would have therefore had absolutely no problem in seeing non-moving objects. The acuity of their eye sight is actually extremely high and rare amongst dinosaurs and actually relatively rare amongst animals in general. In the book this weakness in the T-Rex sight was explained as being another side effect of the frog DNA, but that doesn't make any sense as the vision of the T-Rex is more to do with the shape of the head and positioning of the eye rather than the eye itself.

So… how realistic was Jurassic Park?

Unfortunately for geeks like myself, Jurassic Park was not very scientifically accurate, however it still remains one of my favourite movies. It was the first time that dinosaurs had been shown alongside humans in an almost believable manner. The science might not have stood up to scrutiny but it was close enough to make Jurassic Park seem like it could be real. We might not be capable of creating dinosaurs in a lab, and the imagination of authors and film-makers might have been not very faithful, but that's not what making a movie is about.

Jurassic Park will be remembered primarily by the movie, and as a movies go it was a classic. It had heroes and villains, monsters both good and evil, the evil corporation versus the innocent family, and the classic science versus nature debate which is always popular. It had it all, as well as its fair share of action, adventure, blood and guts and even a bit of romance, I think. So we can let them off with a little bit of lazy science.

*I am basing this statement on one particular episode of The Simpsons, Treehouse of Horrors X - "Desperately Xeeking Xena" in which Lucy Lawless (dressed as Xena the warrior princess) addresses fans at a sci fi convention. Professor Frink quizzes her about inconsistencies in *Xena*, to which she answers each time "a wizard did it."

I realise that I have already used The Simpsons as an example in a previous chapter, and for this I have no excuse. The Simpsons have simply covered a lot of topics in all their years on air. A point echoed in the South Park episode "Simpsons Already Did It" in which the entire episode is dedicated to the character Butters, trying his best not do things that The Simpsons had already done, without much luck.

The Science of being very nice

"Let us try to teach generosity and altruism, because we are born selfish"

Richard Dawkins

Altruism is generally understood as a deliberate act by a person intending to benefit people other than the one carrying out the act. Altruism is an act in which a person puts themselves out, and gains no apparent benefit, so that *another* can gain an apparent benefit. Are you truly altruistic? Well ask yourself this...would you save the life of you brother if it meant that you would certainly die yourself? It's a tough one isn't it?

The scientifically correct answer is No! But I would if it meant saving two brothers or alternatively eight cousins" if you understand why this is the case then you needn't read any further. If you are intrigued, then I shall explain later. Of course, you could save your brother if you wanted to, it just wouldn't be the wisest option.

The simple concept of altruism was a major headache for scientists for quite some time., when the study of evolution was in its infancy. It just didn't make any sense to many of the most intelligent minds in the world. If Darwin's evolutionary theory of the "survival of the fittest" was to work, which it seemed to do so well and so completely, then how could it explain altruism? Well the religious people got right in on it and actually still are (they are always a bit slow) but it didn't take very long for evolutionary biologists to explain altruism and actually add strength to the theory.

In the study of animal behaviour, altruism is considered an unusual trait, as to increase the fitness of another animal whilst decreasing the fitness of one's self doesn't make sense. Altruism doesn't seem to work in our current understanding of evolution, as we operate on a gene-centred view of evolution. Known as the Selfish Gene Theory, coined by Richard Dawkins in the 70's, it is the predominant explanation for the adaptation of living beings through natural selection, which acts through the differential survival of competing genes. By increasing the numbers of the alleles whose phenotypic effects helped to successfully promote their own propagation. This theory basically says that organisms and all of their wonderful and marvelous adaptations to life are merely a genes way of making new genes. Your genes or "immortal coils" have been passed on through the bodies of different organisms since the beginning of time, and will continue to be passed on until the end of days. Unless of course you or one of you ancestors fails to reproduce. It is the soul aim of your genes to make sure that you manage to reproduce and pass on their genetic information to another individual of the same species before you die.

So back to my earlier question, "would you save the life of you brother if it meant that you would certainly die yourself?" with the explanation of the gene-centred view of life do you think that my answer is slightly clearer to you now? No? Ok well maybe I need to further explanation.

The Evolution of Altruism
The most common and widespread form of altruism in nature and is known as reciprocal altruism. It is also possibly the easiest of altruistic behaviours to explain and to understand and it the basis of all research into altruism. Reciprocal altruism is when an individual will selflessly aid another individual without seeking any instant payment at that time. However, it will be expected that if that individual ever finds themselves in the situation of needing help that they will receive it from one of the individual that have been helped in the past. The system only works if everyone helps each other and will break down if anyone acts selfishly too often.

The study of altruism was the initial impetus behind George R Price's development of the increasingly well-known Price Equation. The Price Equation has been described as one of the important equations in all of science. He derived his equation whilst studying the altruistic behaviors of slime-moulds. These particular protists live as individual amoebae until they starve, at which point they join together and form a multi-cellular fruiting body. At this point some cells sacrifice themselves to promote the survival of other cells in the fruiting body therefore displaying altruistic behavior. Price argued that although they did form to a single fruiting body they did display altruism because the cells also have the individual ability to reproduce.

The Equation that price was able to come up with out of this was a simple statement about change. It shows how any characteristic of an organism such as body weight, tooth size, proclivity to altruism etc, changes in one generation to the next. He called this character (z), number of offspring (w) and the discrepancy between the character values of itself and its offspring as (Δz), and showed that the change in the population average value of the character between parent and offspring generations is:

$$\Delta z = \text{cov}(w/w, z) + E((w/w)\Delta z)$$

Or, when simplified

$$w\Delta z = \text{cov}(w1, z1)$$

When altruism was entered into the Price equation as a character, the conclusion was that the more altruistic individuals will lose out to the less altruistic. His equation concluded that for altruism to survive in nature there must be an average spread of altruism throughout a group. He proves this through complex jiggling of mathematics. Mathematics is not my strong point so I am going to avoid going any further into the Price Equation other than to say that it helped a lot of scientists better understand the science of altruism.

The findings of price's equations have been verified continuously in research and it has been verified that a certain level of niceness is beneficial, yet too much opens niceness allows the animal vulnerable to exploitation.

As I explained earlier the system is reliant on an average spread of altruism throughout a group, which for a while lead to some people falsely believing that animals were acting for the good of the group, instead of the good of the individual as the Selfish Gene Theory suggests.

Referred to as Group Selection, is a now more or less defunct idea that organisms act in the best interests of the group, and not in the best interest of themselves. Often altruistic behaviour can give the impression that organisms might be acting towards helping each other out, but in all cases they are in fact gaining something from it that will benefit them in the long run. It might not be instantly obvious but no organism will act altruistically without a decent explanation as its genes will not allow it.

As I have tried to explain, altruism is pretty common in nature if you are very closely related. But what about animals that aren't related, how common is altruism then and how do you explain it?

The most common and best example is in the vampire bat. They live in huge colonies and have to consume blood meals every night, or there is a real chance of starvation. It's hard to find enough blood every night though, so they come home with empty bellies sometimes. And this is when it gets interesting.

The bats start to beg with each other for food. Now these bats aren't related and have are under no genetic obligation to give each other any food. As a result, quite often the bats beggars will be completely ignored and sent packing. And why not? Why sacrifice some of your hard-earned dinner for no good reason? It's of none of your concern whether a none related bat lives or dies. Or is it?

Research has shown that the bats can remember which bat gave them food and which didn't. Also that bats that were given food always reciprocate and return the favour to bats that fed them. Alternatively, bats that did not feed find it harder to get fed. So it pays to be nice and may even save their life. It is easy to see how this strategy could become a stable one but as Prices Equation suggested, you can see how this strategy could become vulnerable. Individuals may stop going out to feed at all and may just rely on all the nice bats around the cave. If too many bats did this the entire system may collapse, so it's important to the survival of the individual to be nice. So the bats only act nice to improve their own chances. Selfishness being the driving force.

Conclusion

Altruism is not a phenomenon that is hard to understand, just as a man might adopt the children of a new wife's ex-partner. He has no genetic obligation to these children (as he isn't related to them) however he may have one in the future with their mother. So it is a smart move to behave altruistically to become as close to the mother as possible. The best way to do this is through her children, as a mother love is unconditional*. Before you judge the father for being cold and calculating, this is not a conscious decision but a pre-programmed and instinctual action. What might on the outside initially appear to be selflessness, when explained further, appears to be selfish however it is not really either. It is merely our genes exploiting our behaviour to reproduce themselves.

So...would you save the life of you brother if it meant that you would certainly die yourself? Now I hope you understand why the correct answer, via the laws of the nature is "No". Unless it were two brothers or eight cousins, as you have the same level of genetic investment in them as in yourself. Conversely the answer for me is "Yes," as I have an identical twin brother. So saving his live in sacrifice of my own would be worth it genetically speaking. For those that haven't worked it out, you have two parents and get half you DNA from each therefore you share half your DNA with each brother or sister you have and an eighth with each cousin. Therefore, to justify wasting your own genetic line, you would have to make up for it by making sure that enough of your own DNA survived to be passed on through relatives by saving enough of them.

Of course I'm not suggesting that this is what would happen, people aren't setting around working out their genetic investment in different relatives. However, this is the whole driving force behind life and the whole driving force behind nice. People act nice so that people think they are nice, they do good things so that people think they are good people and they treat people well so that other people will hopefully treat them well too.

I could go on for ever about different examples of altruistic behaviours in humans and other species and I'm certain all have an explanation. Luckily for us most people are good people and most people do act nicely towards one another when the opportunity arises. People who try and take advantage generally lose friends and learn quickly that it pays to be nice to each other, or at least not to be nasty.

What is Schrödinger's cat?

"This unusual animal (so it is said)
Is simultaneously live and dead!
What I don't understand is just why he
Can't be one or other, unquestionably"

Cecil Adams, 1982

The above quotation is part of a long poem by Cecil Adams, the columnist billed as the "Worlds Smartest Human" in the Chicago Reader newspaper. His question and answer column has been published since 1973 and is particularly renowned for his abrasive humour and incisive wit. The poem is structured as a question by a reader, of which Cecil then answers with apparent ease.

Whether or not Cecil Adams really is as smart as we might be led to believe I don't know, but I certainly couldn't describe the ideas of Schrödinger's cat to you in a poem. As a scientist, I have always had an unspoken awe for theoretical physicists. I think it may be derived from the fact that as I biologist, theoretical physics is about as far away from my field as you get. Theoretical physicist's way of thinking is so different from the majority, that almost nothing they talk about makes much sense to anyone other than theoretical physicists. To be honest I have been trying to teach myself about theoretical physics for a few years now, and I still don't understand much of it.
Schrödinger's cat however is a very simple thought experiment and is easy to understand, but the theory behind it is not quite so easy to get your head around. So I am almost as much writing this chapter to help myself understand it as to help you. As a result, this chapter will either be very long and boring or very short and unsatisfying. Where have I heard that before?

So what was Schrödinger's cat?
There never actually existed a Schrödinger's cat, literally speaking. It is possible that the Austrian physicist Erwin Schrödinger, who devised this thought experiment, may have had cat, however this experiment as never actually been conducted and was never intended to be conducted.

The idea of a thought experiment is to explore and push the potential outcomes of a principle of a certain experiment, which may or may not be possible to perform. Basically they try to "push the envelope" of an idea to its furthest limits with the hope that it will allow the researchers working in this field to either confirm their ideas on their work, challenge their idea or re-think their own work so they can ultimately strengthen it. Basically, thought experiments are designed to show weaknesses in other people's ideas. The Schrödinger's cat thought experiment was designed as a response to what Schrödinger saw as a problem in the then new theory of quantum mechanics being applied to everyday objects, and the strange nature of quantum super-positions.

Quantum super-positions imply a combination of all possible states. Simply put, a single entity exists in more than one state at one time, and according to quantum mechanics it is true. Quantum mechanics also says that at the exact moment that the state of the nature of the position of the matter is measured, it collapses into one definite state. All sounds a little bit crazy doesn't it?

What Schrödinger did was made people re-think quantum super-positions by putting a face to it, well a cat. By describing quantum physics into the real life situation he enabled people to picture it in a living scenario.

Schrödinger's cat, describes a cat being locked up inside a metal box, which must be secured from any type of interference. Inside the box there was a Geiger counter, with a tiny amount of radioactive material attached. If the atom decays it will be measured on the Geiger counter, which will trigger a small hammer to release, shattering a small flask of hydrochloric acid, which will kill the cat.

There is no way to know if and when the radioactive material will decay. If you leave the experiment for a short time there is a fair chance that the cat is alive, but you won't know until you look. If you wait longer there is a good chance that the atom will have decayed and the cat will be dead, however you can't be sure until you look. According to quantum mechanics the cat becomes transformed into "macroscopic indeterminacy" which can only be resolved by direct observation. This is Schrödinger's explanation of quantum super-positions being observed and collapsing into one definite state.

What Schrödinger is trying to do is to create an interpretation of quantum mechanics that will allow us to better understand how it might affect everyday life. The ultimate aim of the experiment however, is to pose the question of when does a quantum system (the cat) stop existing as a mixture of states (alive and dead) to become one or the other?

If the atom did not decay and the cat did not die then the cat remembers only being alive; however as far as we are concerned the fact that no one could observe the cat alive still means that it was both alive and dead, until it is literally observed alive. This is where the paradox lies, it is impossible to know whether the cat is alive or dead and in the very act of checking, it is the observer which decides whether our subject is a living or late cat. Until the observer collapses the wave function, the cat is both alive and dead. The act of interfering will cause the indeterminacy to choose a path, and become determinate.

I know it seem like quantum mechanics is very far-fetched and even ridiculous, but I can assure you it is scientific fact. Come on, you didn't actually think that the harder scientists looked the simpler things were going to get. The exact opposite is the case, particle physicists often joke that they are going to stop looking for sub-atomic particles soon because the more they look for sub atomic particles (particles smaller that an atom) the more they keep finding and its getting very confusing and a bit tiresome. I can understand why you might, after taking the time to read this chapter be asking yourself what use is an experiment that hasn't even been carried out? Schrödinger's experiment was vital in allowing scientist to better their understanding of the developing theory of quantum physics. This was vital as quantum theory was to become one of the most important fields of scientific understanding in the history of science. It is also important practically, as quantum mechanics doesn't just help us understand our universe. Without an understanding of quantum effects we wouldn't have inventions such as computers, lasers and electron microscopes. That would mean CDs and DVDs wouldn't exist, but we also would have no understanding of DNA and no genetic engineering.

Possibly most exiting application of quantum theory in the future is the quantum computer, which will use quantum super-positions to perform infinite calculations instantly by dividing into infinite versions of its self in separate quantum states. It is easily possible that the quantum computer will be able to perform more calculations in an instant, than there are particles in the universe.The quantum computer sounds like something of science fiction and currently it is, however there is rapidly progressing research in that field. Currently it is estimated that in 40 years or so quantum computers will possibly replace the home PC. Hopefully we may be able to look back on this chapter in 40 years with a smile.

Conclusion

I could definitely have explained Schrodinger's Cat a lot more fully and in much more detail but I chose not to, and I tried not to go off on too many tangents, which is difficult when talking about something so complex. I hope you at least understand it a little, has it is a pretty common topic of reference in popular culture. For example, Schrödinger's cat has been mentioned in TV shows such Monty Pythons Flying Circus, Bones, Futurama, The Big Bang Theory, CSI, Doctor Who, Six Feet Under, Star Gate SG-1, Sliders, West Wing, The Sopranos and Yu-Gi-Oh! So now you won't miss those vague references if you ever come across them, useful eh?

How to become a Fossil

"As thou hast given him power over all flesh, that he should give eternal life to as many as thou hast given him"

John 17:2

Obviously immortality is impossible, in the conventional sense. However, if you did want to attempt immortality in a less conventional manner, it may be possible. If you wish to last the ravages of geological times, then the only way to go would be to try and make yourself into a fossil.

Unfortunately, being a human being you have made a poor start. We humans are quite fragile and have small, thin and easily broken bodies. This is why it's usually just our teeth that archaeologists are able to recover from digs.

You would have the best chance of producing a fossil if you had a hard exoskeleton and lived underwater. Your chances of becoming a fossil are significantly improved if you live in sea. Even more likely if the water is very deep, as the shallow water is full of life that will be only too happy to consume your remains. Yet in the deep sea there are much fewer creatures per area squared, and even less beneath the sea-bed, so getting yourself under there would be your best chance. However, humans don't live in the sea, and don't have a hard exoskeleton, so how can a human increase their chances of becoming a fossil on dry land?

Being that us humans are made of mainly soft body parts and live on dry land, the best chance we have is to concentrate on the preservation of our teeth and bones. If you want these body parts to become preserved, you might do well to concentrate on building up your bone calcium levels by eating plenty of cheese and drinking plenty of milk. As your teeth have the best chance of surviving the fossilization process, these are your best chance for a long term survival. So get yourself a quality dentist and keep your appointments!

Now you have done most that you can do to improve your body, the next most important thing comes down to location. You need to try and find a nice place to die that is not going to be disturbed for a long period of time. This is important as you don't want your body to be found, eaten, dismembered and spread-out by scavengers or broken down by insects and bacteria. Caves might be good for this as they are pretty scarce of life and as a result caves usually turn out to be very good sights for the discovery of fossil remains. Taking up potholing might be a good idea, to give you a chance to scout-out some decent places to die, as you will need to go in very far to find a place deep enough that it will be free of life.

If you really want a proper fossilization burial, a rapid burial is the way to go. By a rapid burial I mean a burial that is dramatic and natural, possibly something like a mud-slide or desert wadi during flash-flood season. The sort of conditions that you are really looking for are to be buried in very fine, anoxic mud. Alternatively take a picnic of the flanks of an active volcano, but be sure that you find yourself encased in a nice fine ash-fall, not covered in lava. By choosing a rapid burial it would mean that your entire body would be covered extremely quickly and you would be covered up to a depth of several metres. Your death would not be fun, your lungs would fill and you would die by suffocation but your chances of preservation of your entire body (even soft tissue) are higher.

Probably the best technique of having your self-preserved would be to become incarcerated in tree amber. There have been large amounts of insect and even lizards found preserved in tree amber dating back for millions of years. Some of these have been so well preserved that they even contain small fragments of DNA. However, it would be very difficult to gather enough amber to preserve an entire human body, and you would need to somehow arrange to have the amber buried in a highly stable environment, which would be impossible from beyond the grave.

If you haven't worked it out yet, it is of course almost impossible for any animal, never mind us feeble humans, to become a fossil. You have a far better chance of winning the lottery than you do of becoming a fossil. An even then, you are just a rock.

What makes us see stars?

"Like little dots,
Or specks of paint,
Just floating up above.
I wonder,
What they do up there,
And what they are made of."

 Stars by Gareth Lancaster

The technical word for seeing stars is a phosphene, which is from the Greek words for *phos* (light) and *phainein* (to show,) and they are what are known as entopic phenomena. The definition actually goes someway to explain the phenomena of seeing stars as entopic phenomena are defined as visual effects which under suitable conditions on the eye may render objects within the eye as visible. Entopic images are images (like stars) that are cast in front of the eye, the retina, but have a physical basis. They differ from optical illusions as they are constructed by the brain, instead of the eye.

It was the ancient Greeks that were the first to notice that if you rubbed your eyes hard enough and for long enough, you could cause yourself to see something similar to seeing stars. The pressure caused by rubbing and pushing the eyeball, stimulates the cells of the retina, which can induce a range of phosphenic reactions.

Experiences can vary, and include a darkening of the visual field that will move against that action of rubbing, a strange bright patch of colour that also moved against the rubbing, an ever changing and deforming light grid with black spots and intense blue dots of light in a field of nothingness. All of these sound very abstract and strange but if you have even experienced them or do experience them in the future, you will know exactly what I am referring to. These are known as pressure phosphenes, and they will stop pretty much as soon as you stop rubbing your eyes. However, pressure phosphenes are not the topic of this chapter.

The shapes that you see when seeing stars don't really look like stars at all. They aren't very bright, they don't remain stationary for long and you can see them in clear daylight. What you are actually seeing is activity from your primary visual cortex being cast onto your retina. The stars that you see are actually neurons, which are nerve cells of the brain that are misfiring. Usually this wouldn't matter and you wouldn't notice, but the sharp blow to the head has caused a sharp de-acceleration in the movement of brain. The change in speed happens so quickly that the neurons that are closest to capillaries (blood vessels) change before the neurons surrounding them. This causes them to fire off randomly which your brain will interpret into your vision. So basically what you are seeing in front of your eye is over-excited and confused brain cells.

The most common reason for experiencing this pseudoastroneurological phenomenon is actually the simple act of standing up too quickly. As seeing stars is usually caused by a lack of blood reaching the brain and therefore a lack of oxygen. This also accounts for the light-headed feeling that one might also get. It is most common after standing up quickly from sitting or straightening after bending down, as these are the quickest actions which take a person from fully down to fully upright. Luckily you usually don't see stars as the arteries that are connected to the brain dilate in reaction to your movement, which allow them to maintain the correct blood pressure. However, sometimes they can be caught out, particularly if you are especially tired or are carrying an illness or infection.

Conclusion
If like me, seeing stars is something that you have experience all my life and always wondered about, I hope this chapter was interesting to you. That said, you are in the minority, as although people are quite interested in the topic the majority of people have never actually seen stars before. Some didn't even realise it was an actual thing. So if you have seen stars you know why, and if you haven't you also know why.

What is Brain-Freeze?

"Love is like a brain freeze,
Starts out cool then SHOCKS you
with a BURST of flavour,
calms down again
and leaves you wanting more!"

PoetryNerd95 on Yahoo! answers

Living in Scotland as I do, you might not think we get the opportunity to eat much ice cream. Quite the contrary, the UK has an extremely high consumption per capita of ice cream; on average a person eats 6 litres every year. The majority of this is most likely carried out in the occasional week or two per year when it doesn't rain. In the UK, on sunny day's people decide all of a sudden that they care about being outdoors and fresh air is great. And nothing goes down better on a sunny day than an ice cream.

There is unfortunately one common down side to the overexcited indulgence of frozen food stuffs that can sometimes results from this overindulgent excitement. Common enough, that the phenomenon has been the subject of studies conducted by both the British Medical Journal and Scientific American. Children especially are often so keen to get their mouths around their ice cream that they often end up receiving quite a shock (almost literally) to which they weren't expecting.

What is Brain-Freeze?
Brain-Freeze is also known as an "ice cream headache" or more "technically cold-stimulus headache" or to give its proper name (which you can quickly forget) *sphenopalatine ganglioneuralgia.* That basically just means nerve pain (neuralgia) of the sphenopalatine ganglion, if you don't like the name you can blame Johann Friedrich Meckel or give it its alternative name, the Meckel ganglion. Alternatively, you could do neither and not give it another thought because honestly, how many people actually care about a small parasympathetic ganglion in the upper part of the sphenomaxillary fossa?

Anyway, the first important fact about brain freeze is that it has to have all round decent weather to occur; by this I mean that it is impossible to suffer one in very cold weather. So the bigger the difference between the ambient air temperature and the temperature of the ice cream the better, if you want a frozen brain. So brain-freeze results from eating or drinking something cold quickly.

Another important part of receiving brain freeze is that the food or drink should touch the roof of the mouth before swallowing. This is because there are a large number of important nerve endings in the palate that are important for the ice cream headache to occur. After you stuff your mouth full of the cold substance it usually takes around 10 seconds for the headache to appear. The pain appears on the same side as the cold substance touched the palate, however if you swallowed the ice cream (or whatever it was) whole, then you will get brain freeze on both sides of your head, so that would be ill advised as well as disgusting. In both cases the pain doesn't last for too long, and the brain freeze should be over within about 20 seconds or so.

Is your brain really frozen?
Well ...no, that would be silly wouldn't it. What actually happens when you take a huge mouth full of frozen foodstuff is that you force blood vessels to constrict (tighten) and then dilate (open) extremely quickly as they are re-heated by the warm ambient air all around them. So the blood vessels hide when they are touched by ice-cream, then you swallow it and they explode back to full size as the warm air re-heats them. Unfortunately for us, all of these minute actions don't go unnoticed by the aforementioned nerves which are attached to pain receptors. Each dilatation of a blood vessel in the palate is sensed by a pain receptor, which in turn sends a signal to the brain via the trigeminal nerve. This nerve is the major facial nerve and it covers a large part of your face. The trigeminal also senses facial pain, so is important as it interprets signals sent from the forehead about your brain-freeze. I can vouch for the trigeminal nerves ability to sense facial pain, as when I fractured my face playing rugby it was very effective, it doesn't work as well now however.

Conclusion
So basically as far as I can tell if you get brain-freeze, well for a start you need to stop being so greedy, but basically it is just an unfortunately bi-product of circumstance. I guess it might happen something like this:
It's a beautiful hot day and you are going for a stroll along the beach. The sun is beating down as you stare upward and wipe your brow. As your tired gaze drops, you remove your sun glasses and your eyes re-

focus on an ice cream van. You trot over excitedly and order the largest one they have on the chart. You can't hold back your excitement and you bury your face in its delicious frozen goodness, getting a mouth full of ice cream and an ice cream goatee to match. You're in heaven! You swallow it down and you're just about to have a second, more subtle "lick" when it hits you....ARGHHH!! BRAIN-FREEZE!!

Your blood vessels have already contracted and explosively dilated, sending a panicked and grossly exaggerated message of pain through your trigeminal nerve, straight to your brain causing pain in your sphenopalatine gangalion. For the next 20 seconds you will be holding your head and writhing around in pain.

Oh well... maybe you should mind your manners the next time you eat ice cream.

The Science of Homeopathy

"In homeopathy, you'll be a hero, if you can prove that zero isn't really zero."

Kieran (twitter blogger)

Homeopathy seems to be one of the only branches of alternative medicine that continues to increase its following over time and trial, regardless of trends and fashions. Advocates of homeopathy site this as the greatest proof of its effectiveness; as if it didn't work people would stop using it. I know many people who swear by its effectiveness and some even go so far as to state it is superior to modern medicine. Indeed, homeopathy has been funded by the NHS and there are at least five homeopathic hospitals with homeopathic doctors in the UK.

Taking this into consideration it might be easy to become sold on the value of homeopathy as a legitimate science. That is until you actually take a brief look at the actual science behind homeopathy.

Homeopathy dates back some 2500 years to the Greek physician, Hypocrites. Hypocrites argued that "By similar substances a disease arises and by administering similar things they regain their health from sickness". Sounds ridiculous, until you consider that this is the idea behind vaccinations. Not entirely surprising then that a German Physician named Samuel Hahnemann, decided he would try and see if the principle worked for non-infectious disease as well.

The first problem Hahnemann found however that was too high a dose was a very bad thing, and tended to make things much worse. He therefore deduced that he must dilute his remedies, and this is when he put Homeopathy into direct conflict with science. Homeopathy stated that *the greater the amount of the dilution, the stronger the remedy.*

This idea is more than a paradox; in many remedies the dilution is so great that not even a single molecule of the original extract remains. Some of the homeopathic dilutions on sale today are diluted on a comparable level as one grain of sand, in all of the desserts and beaches on the planet. Hardly potent I'm sure you will agree. Homeopaths argue that this is not relevant as the water that the extract is diluted in has 'memory' and stores all the information from the original extract remains within it. I'm not sure whether it is the Hydrogen or the Oxygen that has the brain however. As far as I was aware water only contained two atoms of hydrogen which is covalently bonded to one atom of oxygen, no brain.

As a scientist it is not fair to deny the effectiveness of a treatment just because it sounds crazy. For example, the mechanisms of anaesthesia remain poorly understood, however I doubt anyone would deny its effectiveness. Does homeopathy work too?

Does it work?
The fact of the matter is that currently there is no evidence to support the effectiveness of homeopathy. There have been a large number of studies into the effectiveness of homeopathy, the most convincing one concluding that the results were encouraging but not convincing. It Concluded that there was insufficient evidence to support homeopathic therapy. The majority of studies found the effectiveness of homeopathy to be on a par with the effectiveness of the blank sugar pills used as a control in the experiments. Homeopathy is essentially nothing more than a placebo in a fancy dinner jacket.

How is it meant to work?
The reality of the matter is that homeopathy goes far beyond any current understandings of physics or chemistry. In truth, going on our current understandings of science, homeopathy is literally impossible, something scientists don't like to say of anything. This is mainly due to the enormous dilutions used in treatments, and by the fact that higher dilutions are said to be more potent. Physicist Robert L Park, former executive director of the American Physical Association, has noted that since the least amount of a substance in a solution is one molecule, a 30X solution would have to have at least one molecule of the original substance dissolved in a minimum 1 billion billion billion billion billion billion billion billion molecules of water. This would require a container more than 30 billion times the size of the Earth. Park has also noted that if you actually wanted to guarantee that you would receive even one molecule of the original substance allegedly present in 30X pills, it would be necessary to swallow 2 billion of them, a total about 1000 tons of lactose plus whatever impurities the lactose contained. Basically, it's laughably ridiculous.

So why do people use it?

But let's not get all supercilious, just because we have our constant allies of science and logic on our side. Millions of people all over the world swear by this stuff. Why? This is an interesting question, and is really a lot more to do with psychology than the effectiveness of the treatment. There are many reasons why a treatment might seem to be effective and I will not bore you with them all, however luckily the simplest to explain are generally the most common.

Firstly, the majority of people who take homeopathic pills take them for minor illnesses and ailments, things that do not last for very long and would have cleared up, in time, on their own. Your body has a remarkable ability to cure itself and homeopathic pills seem to be taking much of the credit for your body's natural healing powers. They may take homeopathic pills for a while, get better on their own and thank the homeopathy for it.

Secondly and along similar lines, when people become ill they often take a mixture of different potential cures with the hope that at least one of them will make them feel better. For example, if a person has a cold, they might go to bed with a hot lemon drink, after consuming some pain-killers, cough medicine and throat sweets. The next day feeling much better, and enthusiastically declaring that the hot lemon drink was a miracle cure. Who don't know that it was the hot drink; it could just as easily have been one of the other treatments or most likely, the good long sleep and decent amount of rest that she had just enjoyed. Homeopathic remedies can benefit from similar mistaken identity, especially if it happens to be the first time they have used it. All of a sudden it is the pill and not all the other action taken over the course of the illness which has cured them. This is hard to argue against or for as it is difficult to prove or deny. Most people are naturally positive and trusting and like to believe that things are working, even when they aren't. The placebo effect is another very likely possibility for the false positives people sometimes think they are experiencing. This is a well-studied and poorly understood phenomenon. Studies have shown that some people can become cured by a pill which they believe to be a drug, even when it isn't. It is possible that some people experience this affect just because they think the homeopathy is a legitimate treatment. That said, the bulk of research has shown the effectiveness of homeopathy to be lower than that of a placebo, which doesn't add much strength to the argument for homeopathy.

There is a great deal of other reason why homeopathy is quite ridiculous. Some are so crazy that I am actually slightly embarrassed to even mention them. Such as I mentioned earlier, the idea that homeopathy is reliant on the unproven and pretty silly idea that water has a memory, and the ability to store chemical memory within it. This begs the question, what about the clouds the water came from, the mountain streams that it flowed down or the metal piping it travelled up. I can only hope it only has a short-term memory because surely not all the apparent memories, from that epic journey can't be pleasant or healthy for human consumption.

In closing
I actually know several people who are very involved in the use of homeopathy. I have questioned them several times about the origins of homeopathy, the crazy dilutions and several of the other things that make it seem crazy. It seems that they have forgotten the roots, others don't believe a lot of the original truths that homeopathy was founded on. Some are even trying their best to tie it in with the modern fields of electromagnetism, and even quantum physics. I guess if they can confuse their patients with the apparent science, it makes homeopathy seem more plausible. They are possibly aiming the use the pseudoscientific tactic of baffling the lay-person in bad science and miss-information, some that you might have to be a scientist to know was nonsense. Sometimes I wonder if the homeopaths really think that they are talking good science too, as many of them are just reading books, attending courses and living off information handed down to them. Are they suckers too? Homeopathy is one of those things that most people don't actually understand what it is. In my own small amount of research, I have found that most people believe that homeopathy is some kind of herbal remedy. I'm pretty certain that if the majority of people that used it were aware of exactly what it was, they would be less willing to part with their hard earned cash on what is essentially nothing but a blank pill.

Can you die of a Broken Heart?

"It hurts to breathe because every breath I take proves I can't live without you."

Anonymous

Anyone who has lost a loved one will empathise with the sentiments expressed by poor old Anonymous, in the above quotation. I'm sure he or she was thoroughly heart broken when they made this statement. The passion and emotion expressed in this one sentence can be understood, and felt. You can understand his/her pain.

The broken heart is a common metaphor, used to describe a sense of indescribable pain felt when one suffers severe emotional loss, which seem manifests in actual physical pain. Sometimes is it referred to as heart-ache, and that is very much the way it feels.

The broken heart feeling is also thought to be closely related to that sinking feeling that you get when you are nervous, or under stress. However, the broken heart feeling lasts much longer.

The truth is that severe emotional damage can actually cause damage to a healthy heart! The odd and fascinating phenomenon known as "Broken heart syndrome," which I will come to later in the chapter.

How to spot a broken heart
You always know when something terrible has happened to you, but do you really have a broken heart, metaphorically and literally? There is a long list of symptoms that can occur as a result of a broken heart. The first physical symptom, the tightness in the chest which is similar to an anxiety attack. Other related physical symptoms include sickness, loss of appetite and tiredness. A broken heart can also have serious psychological impacts on a person, and can lead to minor affects such as crying and apathy. In some severe cases, extreme feelings of despair or suicidal thoughts can be felt by the broken hearted.

Broken heart syndrome
Broken heart syndrome goes under a lot of different names, it can be referred to as Takotsubo cardiomyopathy, transient apical ballooning, apical ballooning cardiomyopathy, stress-induced cardiomyopathy, broken-heart-syndrome and the most common ... stress cardiomyopathy. Takotsubo cardiomyopathy, is the original name given to the condition, earned from a type of octopus's trap

from Japan where the condition was first identified. Typical of Japanese legends, the story is romantic, poetic and full of heart-break but also slightly silly. It tells of a fisherman who fell in love with an octopus in which he caught, however she did not love him back and he died of a broken heart.

The most common name given to broken heart syndrome however is stress cardiomyopathy, and as with all medical conditions you can work out what the symptoms are by the name. Cardio is a reference to the heart, Myopathy means any or all type of disease or damage to the muscle. So stress cardiomyopathy means a stress induced damage to the heart muscle, the stress part coming from the trauma of heartbreak.

For medical purposes a reference to stress is a reference to anything that is not normal. These abnormalities can be a massive range of physical and emotional things such as body temperature, emotional state, low blood sugar, dehydration etc. When these abnormalities occur, your body responds by producing various hormones and proteins to help you cope with the stress and deal with it better. This is where it is believed that the problems begin with stress cardiomyopathy. When the body undergoes a large amount of stress it produces a great deal of one particular hormone - adrenaline. Adrenaline is a vital hormone however it can also be a killer. In stress cardiomyopathy, it is believed that the heart muscle is overwhelmed by the massive amount of adrenaline that is suddenly produced in response due to high stress levels on the body.It seems as if the heart muscle is almost stunned by the chemical influx of adrenaline, and in response will almost cramp up and stops working. Also like a cramp, there is no permanent damage and unlike with a heart attack the muscle cells are not destroyed. This stunning gets better after only a few weeks, even if a person was suffering from severe muscle weakness when taken to hospital, full recovery with no permanent damage is likely in just about every cases.

Can you die of a broken heart?
Stress Cardiomyopathy is not fatal by any means; it may feel like very much a heart attack, but broken heart syndrome only seems to temporally break your heart. However, what if your love was so deep, so long lasting and meaningful that it was everything that you lived for. Could the loss of a love that deep kill you?

Well consider this. I had a great grandfather called Jimmy, a wonderfully fascinating man. He was in two World Wars and was a

Japanese POW in WWII. Brought up in Singapore in a rich British family, the truth of his family background is hazy. He and his wife Dye were married forever, and lived together in Dorset. They were a wonderful couple, when I met him he was approaching 100, and she was in her 90'. However, a few years later Dye died, a very short time afterward so too did Jimmy with no explanation at all. His local doctor said that he had "just had enough of living." Could it be that Papa Jimmy died of a broken heart? I have never thought about it until now. However, after such an incredible life full of drama, strife, love and turmoil, I couldn't think of a more poetically beautiful way to end it.

In closing

"There is grandeur in this view of life, with its several powers, having been originally breathed by the Creator into a few forms or into one ... from so simple a beginning endless forms most beautiful and most wonderful have been, and are being evolved."

This is part of the last sentence of "The Origin of Species..." by Charles Darwin and I think it is one of the most beautiful and elegant pieces of writing I have ever read. Never before has a single sentence meant so much to me and had such an impact on my life, except the creator part and to be honest that didn't mean much to Darwin either. I bring it up firstly because I have just had the first part "There is grandeur in this view of life" incorporated into a new tattoo, emblazoned on banners surrounding a Galapagos Finch. If you know your Darwin, you will know the significance of the Galapagos finch. Secondly, I mention it because the way Darwin ended his masterpiece is the way in which I would have like to have ended my book.

Alas this is not a book of anywhere near the same significance and I am not that a writer or scientist of anything like Darwin's talents. Most significantly, this is a very different time in which we live.

Science is about discovery, as it always has been, but unlike in Darwin's day when many discoveries were being made for the first time today's scientific breakthroughs are much more "provisional." It isn't that we don't make as many or that the scientists are as good (the exact opposite actually) but that we except that when new discoveries are made, that they are one small steps closer towards a betting understanding. It is understood the better machines, better techniques, better scientists will slightly modify the body of knowledge in due time.

This has always been part of science and it is happening faster now than it even has. It is critical to the advancement of science and is the refinement of the way in which we understand the universe in which we live. It strengthens science by making never allowing it to stagnate and constantly allowing the wheel of knowledge to gather momentum.

It is often a criticism of people out with the scientific community that science contradicts itself, or that scientists are always changing their minds about key issues. They site this as a weakness, stating that it must mean that ALL science therefore must be wrong. The truth is that even if scientists were like that, denying ALL science still wouldn't make sense. Scientists only appear to contradict each other because science isn't always as black and white as the mass media would like it to be. Massive disagreements do not occur often. Yet the illusion can be created as new studies do a slightly better job than older studies ... Oh the drama!!

Sometimes science suffers bad press. Scientists are stereotyped for being geeky, awkward, unattractive to the opposite sex, unsocial, unfashionable, short-sighted (the works) for no good reason. A lot of scientists I have met do have all of these features, yet so do a lot of non-scientists. Always the scapegoat for the world's problems, it is also the scientists of the world that we turn to for all of the answers also. Science is a wonderful thing, easily the greatest creation of the human mind and I hope that I helped you to see it for what it is.

Cheers